CONFRONTING SPACE DEBRIS

STRATEGIES AND WARNINGS FROM COMPARABLE EXAMPLES INCLUDING DEEPWATER HORIZON

Dave Baiocchi • William Welser IV

Prepared for the Defense Advanced Research Projects Agency

RAND | NATIONAL DEFENSE RESEARCH INSTITUTE

The research described in this report was prepared for the Defense Advanced Research Projects Agency. The research was conducted within the RAND National Defense Research Institute, a federally funded research and development center sponsored by the Office of the Secretary of Defense, the Joint Staff, the Unified Combatant Commands, the Navy, the Marine Corps, the defense agencies, and the defense Intelligence Community under Contract W74V8H-06-C-0002.

Library of Congress Control Number: 2010940008

ISBN: 978-0-8330-5056-4

Published 2010 by the RAND Corporation
1776 Main Street, P.O. Box 2138, Santa Monica, CA 90407-2138
1200 South Hayes Street, Arlington, VA 22202-5050
4570 Fifth Avenue, Suite 600, Pittsburgh, PA 15213-2665
RAND URL: http://www.rand.org/
To order RAND documents or to obtain additional information, contact
Distribution Services: Telephone: (310) 451-7002;
Fax: (310) 451-6915; Email: order@rand.org

Preface

Orbital (space) debris represents a growing threat to the operation of man-made objects in space.[1] According to Nick Johnson, the National Aeronautics and Space Administration's (NASA's) chief scientist for orbital debris, "[T]he current orbital debris environment poses a real, albeit low level, threat to the operation of spacecraft" in both low earth orbit (LEO) and geosynchronous orbit (GEO) (Johnson, 2010). There are currently hundreds of thousands of objects greater than one centimeter in diameter in Earth's orbit. The collision of any one of these objects with an operational satellite would cause catastrophic failure of that satellite.

This monograph presents a new way of thinking about the orbital debris problem. It should be of interest to space-faring nation-states and commercial firms, the legislative and executive branches of the U.S. government, the United Nations Committee on the Peaceful Uses of Outer Space, and the general public.

This research was sponsored by the Defense Advanced Research Projects Agency (DARPA) and conducted within the Acquisition and Technology Policy Center of the RAND National Defense Research Institute, a federally funded research and development center sponsored by the Office of the Secretary of Defense, the Joint Staff, the Unified Combatant Commands, the Navy, the Marine Corps, the defense agencies, and the defense Intelligence Community.

---

[1] NASA defines orbital debris as "artificial objects, including derelict spacecraft and spent launch vehicle orbital stages, left in orbit which no longer serve a useful purpose" (NASA-Handbook 8719.14, 2008).

For more information on the RAND Acquisition and Technology Policy Center, see http://www.rand.org/nsrd/about/atp.html or contact the Director (contact information is provided on the web page).

Contents

Figures

Tables

Summary

Background and Objective

Orbital (space) debris represents a growing threat to the operation of man-made objects in space.[2] According to Nick Johnson, NASA's chief scientist for orbital debris, "[T]he current orbital debris environment poses a real, albeit low level, threat to the operation of spacecraft" in both LEO and GEO (Johnson, 2010). There are currently hundreds of thousands of objects greater than one centimeter in diameter in Earth's orbit. The collision of any one of these objects with an operational satellite would cause catastrophic failure of that satellite.

DARPA, within the context of the Catcher's Mitt study, is in the preliminary stages of investigating potential technical solutions for remediating debris.[3] This investigation is a critical step because even the most rudimentary cleanup techniques will require significant research and field testing before they can be successfully implemented. In addition, future pathfinder missions will require extensive resources,

---

[2] NASA defines orbital debris as "artificial objects, including derelict spacecraft and spent launch vehicle orbital stages, left in orbit which no longer serve a useful purpose" (NASA-Handbook 8719.14, 2008).

[3] The DARPA Catcher's Mitt study is tasked with the following objectives: model the space debris problem and its future growth; determine which class of satellites is most affected; and, if appropriate, explore technically feasible solutions for debris removal. DARPA intends to use the results of the Catcher's Mitt study to determine if they should invest in a space debris remediation program (Jones, undated).

and the U.S. government will need sufficient justification before pursuing these programs.[4]

With this background in mind, this research had three primary goals. The first was to determine whether analogous problems from outside the aerospace industry exist that are comparable to space debris. Assuming that such problems exist, the second goal was to develop a list of identifying characteristics along with an associated framework that could be used to describe all of these problems, including debris. The final goal, provided that the first two were possible, was to use this framework to draw comparisons between orbital debris and the analogous problems. Ultimately, we hoped to provide context and insight for decisionmakers by asking the following question: How have other industries approached *their* "orbital debris–like" problems? What lessons can be learned from these cases before proceeding with mitigation or remediation measures?

Comparable Problems

We identified a set of comparable problems that share similarities with orbital debris and narrowed this set down to the following nine issues: acid rain, airline security, asbestos, chlorofluorocarbons (CFCs), hazardous waste, oil spills, radon, spam, and U.S. border control.[5]

These problems are related because they all share the following set of characteristics:

- Behavioral norms (past and/or present) do not address the problem in a satisfactory manner.
- If the problem is ignored, the risk of collateral damage will be significant.
- There will always be an endless supply of "rule-breakers."

---

[4]   Within the scope of this document, we define the word *pathfinder* to mean an experimental prototype used to prove a capability.

[5]   We do not describe the rationale behind this statement in the executive summary. However, more information about the comparable problems is available in Chapter Three and in Appendixes A and B.

- The problem will likely never be considered solved because the root cause is difficult to eliminate.

Nomenclature

We refer to the terms *mitigation* and *remediation* throughout this analysis, so it is important to provide our definitions for these terms:

- *Mitigation* refers to a class of actions designed to lessen the pain or reduce the severity of a problem. Mitigation measures are inherently preventive, and they are enacted to prevent a problem or to prevent one from getting worse.
- *Remediation* refers to the act of applying a remedy in order to reverse events or stop undesired effects. Remedies are targeted reactions often designed to address an undesirable event that has already occurred.

Methodology

We used a literature survey and interviews with experts to gather the following pieces of information for each of the comparable problems:

1. **Basic overview.** What is the problem?
2. **Calendar dates of key milestones.** When was the problem first identified? When were major mitigation measures imposed? When (if at all) were remedies fielded?
3. **Stakeholder demographics.** Who is viewed as having caused the problem? Who is affected by it? How large is each group? How diverse are their interests?
4. **Current status.** What was the status of each of the problems, as of May 2010? Was it being remedied or simply mitigated?

The Framework

Once we had this information, we designed a framework that could be used to describe the process for addressing orbital debris and any of the comparable problems. We identified four stages of increasingly aggressive measures that could be used to address the various problems: identifying, characterizing, and bounding the problem; establishing normative behaviors; mitigation; and remediation.

These stages can be represented with a series of concentric rings, as shown in Figure S.1. This concentric geometry highlights an important feature of the approach: As the community moves toward the center (which indicates increasingly aggressive deterrents), the size of the risk-generating population decreases with each inward step.

The progression through these stages is determined by the risk tolerance of the affected entities. Specifically, decisionmakers should proceed to the next stage when the existing population of unwanted incidents exceeds the community's risk tolerance level. For example, catastrophes—such as an oil spill—can cause a community to reassess (and often lower) its risk tolerance, and additional mitigation or remediation strategies may be needed after such an event.

Figure S.1
Framework Stages via Concentric Rings

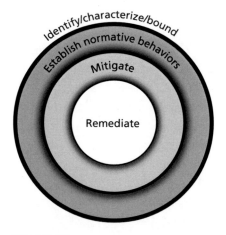

It is also important to note that eliminating the problem is not necessarily the primary objective. Instead, the goal should be reducing the risk posed by unwanted phenomena (air pollution, radon levels, aircraft hijackings) to a level that the affected stakeholders find acceptable.

We also developed two tools to aid in describing the stakeholder communities. These tools are shown in Figures S.2 and S.3, and more information is provided in Chapter Seven.

Figure S.2
Stakeholder Diversity and Type

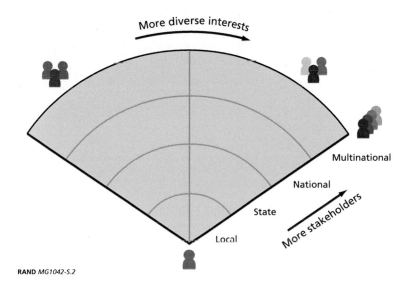

RAND *MG1042-S.2*

Analysis: Comparing the Relative Time Spent in Each Stage

Our literature survey uncovered several important dates, milestones, and achievements associated with all of the comparable problems. We used this information to build a series of timelines that allowed us to compare the different problems. After reviewing these timelines, we made the following observations:

Figure S.3
Stakeholder Spectrum: Blameworthy Versus Affected

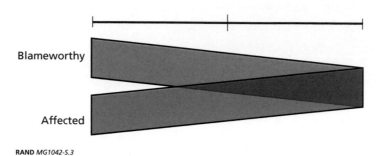

RAND *MG1042-S.3*

- It may take several years to identify the problem (acid rain, asbestos, and spam).
- In some cases, a single critical event is enough to propel the problem through several of the stages shown in Figure S.1 at once (airline security, oil spills, radon, and spam).
- A problem need not have existed for a long time before remediation is deemed necessary (hazardous waste, oil spills, and spam).
- Once in remediation, the problem is not considered solved. Airline security, hazardous waste, oil spills, radon, and spam are all examples of problems that are difficult to completely eliminate.

Mitigation Concepts

We identified three mitigation approaches—taken from the environmental protection industry—that can be applied to any of the problems that we considered, including space debris:

- The command and control (C2) approach institutes an incentive structure to control community behavior. This approach is easily understood by most cultures, so it is often the first mitigation strategy to be implemented.
- Market-based approaches acknowledge that the problem exists and organize a formal allocation scheme for the right to engage in that behavior.

- Performance-based strategies use a quota-based system to set a limit on the undesired behavior.

We highlighted the mitigation strategies used to address acid rain, airline security, and radon. Our analysis of these issues yielded the following observations on each.

Acid Rain

- In order to successfully implement a large C2 strategy, the symptoms must be categorized into groups that represent different levels of relative risk.
- A market approach is often most effective only after an effective C2 strategy is already in place.

Airline Security

- Preparing for potential threats requires an efficient and effective system for collecting and disseminating information.
- An effective mitigation strategy evolves over time.
- A successful C2 strategy is enforced by organizations with clearly defined jurisdictions.

Radon

- Nonregulatory approaches may be good at increasing public awareness, but they are unlikely to achieve high levels of compliance.
- Mitigation is relatively straightforward when the problem can be described and measured accurately.

Remediation Concepts

Types of Remedies

Remedies can be classified using two sets of descriptive categories:

- **Relocation versus elimination.** An undesired object can be relocated such that it no longer poses a high risk, or it can be completely eliminated.
- **Targeted versus dragnet.** Undesired objects can be relocated or eliminated using processes that are either targeted or dragnet-like. Targeted removal techniques use a specific method to affect a single, known object. Dragnet strategies indiscriminately trawl space to gather and remove all objects with a particular set of characteristics.

Lessons Learned From the 2010 Deepwater Horizon Oil Spill

Using this oil spill as a case study, we identified the following lessons learned:

- Simply having one or more remediation technologies is not sufficient. The remedies must be tested and proven to work in the expected operating conditions.
- The community will only support the development of an effective remedy when the risk posed by the threat is considered to be unacceptable.
- When reacting to a catastrophe, a dragnet solution is needed to address the aftereffects.
- After a catastrophe, a targeted solution may also be necessary to remedy a problem.
- Remedies must be adaptable so that they may evolve to face the latest challenges.

Summarizing Observations

We noted the following key themes as we compiled the results from this research:

- Improving situational awareness should be an ongoing effort within any community.

- The Superfund could serve as an effective model for orbital debris cleanup.
- Incentive structures (associated with mitigation strategies) work best in the short term. In order to achieve a cost-effective long-term solution, it is necessary to change stakeholder preferences.
- All of the stages shown in Figure S.1 must continually evolve over time along with the problem.

The Case for Additional Mitigation in Orbital Debris

When viewed in light of the comparable problems, there is evidence to suggest that orbital debris does not at present pose a great-enough risk to warrant the deployment of a remediation technology.[6] A community will only move on to the next stage shown in Figure S.1 when the current stage is not sufficient to properly address the problem. While everyone in the space community certainly agrees that orbital debris poses a risk, the lack of government and private industry funding for this effort suggests that the perception of risk has not yet crossed a critical threshold that would prompt demands for remediation.

The current lack of private funding for debris remedies is particularly telling. Today, the majority ownership of operational space assets (as a percentage of the total operational inventory) has shifted from government to commercial industry.[7] For this new majority of commercial stakeholders, the "imperative to create shareholder value entails that any investment in a technical system be guided by its value creation potential" (Brathwaite and Saleh, 2009). In other words, if debris were deemed to represent an unacceptable risk to current or future operations, a remedy would already have been developed by the private sector.

---

[6] The use of the word *deployment* is intentional: It implies an operational—and not simply a pathfinder—system.

[7] According to the April 2010 Union of Concerned Scientists (UCS) Satellite Database, 41 percent of the world's active, operational satellites are solely commercial; 17 percent are solely military; 18 percent are solely government; and the remaining 24 percent are either multiuse or used for research or scientific purposes. While the UCS database represents only an approximate count of the world's total satellite inventory, it is useful in providing a quick estimate to support our claim (UCS, undated).

The space industry is currently dealing with the debris problem via mitigation, and we offer the following observations about these efforts:

- Mitigation is an effective way to reduce the probability that a catastrophe will occur.
- Tracking metrics over time is an effective way to measure a mitigation strategy's ongoing effectiveness.

The Case for Developing Remediation Technologies

The lack of funding initiatives associated with developing a deployable remedy for orbital debris suggests that the community currently does not need such a capability. However, our research presents several lessons that suggest it may be wise to develop a pathfinder system in the near term:

- **A community must be prepared for "shocks" or catastrophic events.** Sometimes a single catastrophic event, or shock, is sufficient to propel a community through several of the stages at once. For orbital debris, the Chinese antisatellite test and the Iridium/Cosmos collision are two obvious examples (see Chapter One for more detail about these events). These two events are likely the cause for the increased interest—to include this research—in the debris problem. In addition, remedies are needed to clean up the aftereffects of such catastrophic events. Developing the pathfinder technology now for such a remedy may prove to be a wise decision because on-orbit collisions are likely to continue to occur in the future.
- **Remedies must be designed and tested to work under the actual operating conditions.** This is the biggest lesson from the Deepwater Horizon spill. All of the remedies fielded during the first 40 days of the spill were not effective because they had not been tested or proven to work in deepwater drilling conditions. Fielding a demonstration technology will prove useful only if it will provide operators and engineers with relevant information on technology performance under the actual working conditions. In

addition, decisionmakers will gain important data points on realistic values for recharge times, reaction times, and the magazines associated with any potential remediation technology. Ultimately, the pathfinder system must strive toward remedying a realistic problem, or the development will risk being considered purely academic and not operationally useful.

- **One remedy is not good enough.** A remedy is often used to respond to an event that has already occurred. As a result, remediation technology is often very specialized, and our research indicated that for many problems, several different techniques are necessary. There are examples of this throughout all of the comparable problems. Airline security, asbestos, environmental hazards, oil spills, radon, and spam all use multiple techniques to remedy a problem. For this reason, it may be wise to begin developing a pathfinder system now so that alternative, tangential methods may be developed more quickly in the future.

- **When a problem's effects are not directly observable, a community is likely to underestimate the risk posed by the effects.** Asbestos and radon are invisible, and the cancers they cause may not appear for several decades. Under such circumstances, a community may have a low perception of risk because the cause and effect are separated by long spans of time. By contrast, the neighbors of a polluting factory are likely to see its effects every day. Orbital debris, unfortunately, belongs to the category of problems that are not easily observed either by those who create it or by those who might be harmed by it. Because the harm is virtually invisible until a major collision occurs, the broader community may be simply unaware of the severity of the problem, or they may tend to underestimate the potential risk. Therefore, the technical community should consider implementing an ongoing, metric-based stakeholder awareness program alongside the development of a technical remedy.

Acknowledgments

We thank the Defense Advanced Research Projects Agency's Tactical Technology Office, specifically Wade Pulliam and his team, for the opportunity to perform this research on orbital debris. Nicholas Johnson, NASA's chief scientist for orbital debris, played an important role in educating us on current mitigation efforts as well as context for the debris-related timeline and stakeholder risk tolerances. Lt Gen Larry James, 14th AF/CC, and Brig Gen John Hyten, Director, Space Acquisition, Office of the Under Secretary of the Air Force, and their respective staffs provided key background information on the tactical aspects of identifying and managing the existing debris population.

Roseanne Ford, from the University of Virginia, deserves our appreciation for sharing some considerations that affect the development of remediation-focused technologies.

This research would not have been possible without the responsiveness, resourcefulness, and enthusiasm provided by RAND librarian Anita Szafran. We are extremely grateful for her assistance in compiling sources for our literature survey and building the Bibliography and Works Consulted for Timelines lists, which contain over 360 references. Anita embodies the ways in which the library staff is critical to RAND research, and we are very thankful for her contributions to this work.

Sonni Efron reviewed an early draft of the manuscript, and she provided a number of recommendations designed to clarify and refine our conclusions. We are very thankful that she was willing to provide so many thoughtful suggestions.

We are very lucky to have an excellent team of administrative staff that helped us with document preparation and project travel. We have the pleasure of having Holly Johnson and Regina Sandberg support our research activities on a day-to-day basis. Special thanks are due to Karin Suede and Vera Juhasz, who took great care in formatting the document for release to the sponsor and peer reviewers.

We are also grateful for our talented research editor, Nora Spiering, whose expertise in editing and typesetting the final document helped facilitate the publication process.

We thank our colleagues, Natalie Crawford, Tim Bonds, Phil Antón, Kenneth Horn, Karl Mueller, Andrew Morral, and Henry Willis, for their insights and support throughout this project.

Finally, we want to thank our spouses and families for their tolerance and understanding during the three months in which we researched and wrote this monograph.

From the onset we have wholeheartedly believed in the utility and importance of this research and have spent countless hours reading, discussing, and analyzing information on this problem. We have taken extra care to ensure that this document presents an intuitive and accessible approach to the debris problem. We accept the responsibility for the observations and statements within this monograph.

Abbreviations

14th AF/CC	Commander, Fourteenth Air Force
ASAT	antisatellite
C2	command and control
CERCLA	Comprehensive Environmental Response, Compensation, and Liability Act
CFCs	chlorofluorocarbons
DARPA	Defense Advanced Research Projects Agency
DH	Deepwater Horizon
EPA	United States Environmental Protection Agency
GAO	United States Government Accountability Office (formerly United States General Accounting Office)
GEO	geosynchronous orbit
LEO	low earth orbit
NASA	National Aeronautics and Space Administration
pCi/L	picocuries per liter
TSA	Transportation Security Administration
UCS	Union of Concerned Scientists

CHAPTER ONE

Introduction: The Problem of Orbital Debris

What Is Orbital Debris?

Orbital (space) debris represents a growing threat to the operation of man-made objects in space.[1] According to Nick Johnson, the National Aeronautics and Space Administration's (NASA's) chief scientist for orbital debris, "[T]he current orbital debris environment poses a real, albeit low level, threat to the operation of spacecraft" in both low earth orbit (LEO) and geosynchronous orbit (GEO) (Johnson, 2010). This risk poses a threat to the United States' ability to access and use the space environment. For example, on the most recent Hubble Space Telescope repair mission in May 2009, NASA estimated that astronauts faced a 1-in-89 chance of being fatally injured by a piece of debris while operating on the telescope outside the space shuttle (Matthews, 2009).

The United States maintains a catalog for space objects that are larger than about 10 cm in diameter, and this catalog currently contains about 20,000 objects, of which debris constitutes a majority (Kehler, 2010; Space Track, undated). In addition, NASA estimates that there are an additional 500,000 objects between 1 and 10 cm, and that there are likely tens of millions of particles smaller than a centimeter (Orbital Debris Program Office, undated).

---

[1]   NASA defines orbital debris as "artificial objects, including derelict spacecraft and spent launch vehicle orbital stages, left in orbit which no longer serve a useful purpose" (NASA-Handbook 8719.14, 2008).

These smaller objects pose some of the greatest risk to orbiting payloads. As Johnson notes, "[T]he principal threat to space operations is driven by the smaller and much more numerous uncatalogued debris" (Johnson, 2010). In LEO, objects have velocities of 7 or 8 km/s with respect to the ground, which means that even small particles can impart a tremendous amount of energy if they collide with another object. This threat is especially sobering because most small particles are uncataloged.[2]

Prior to 2007, the primary source of orbital debris was explosions of spent rocket engines. Originally, these engines were jettisoned in orbit after launch, and the remaining fuel expanded because of the thermal conditions. Under the right conditions, the pressure became too great, and the rocket body exploded. Since the mid-1990s, engines have been designed with valves that relieve the pressure by venting the residual fuel, and contemporary rocket bodies are no longer a major contributor of debris.

To date, the largest two contributors of debris have been collision events. The first was the 2007 Chinese antisatellite (ASAT) test. As part of this test, China launched a ballistic missile and hit the Fengyun-1C, a defunct Chinese weather satellite. This collision event generated a debris cloud that has added 2,606 trackable objects to the U.S. space catalog as of June 2010 (Space Track, undated). In addition, some estimates suggest that between 35,000 and 500,000 smaller, untrackable pieces of debris were created as a result of this test (Carrico et al., 2008). The second event was an inadvertent collision in February 2009 between an active Iridium communications satellite and Cosmos 2251, a retired Russian communications satellite. This crash added 1,658 trackable objects to the U.S. catalog as of June 2010 (Space Track, undated).

---

[2]   By contrast, larger objects are relatively easy to track and catalog. In addition, many large objects in LEO will eventually fall back to earth. This is because the larger objects have a higher drag coefficient, so they tend to slow down, enter the earth's atmosphere, and burn up upon reentry.

Objective

The Defense Advanced Research Projects Agency (DARPA), within the context of the Catcher's Mitt study, is in the preliminary stages of investigating potential technical solutions for remediating debris.[3] This investigation is a critical step because even the most rudimentary cleanup techniques will require significant research and field testing before they can be successfully implemented. In addition, future pathfinder missions will require extensive resources, and the U.S. government will need sufficient justification before pursuing these programs.[4]

With this background in mind, this research had three primary goals. The first was to determine whether analogous problems from outside the aerospace industry exist that are comparable to the problem of orbital debris. Assuming that such problems exist, the second goal was to develop a list of identifying characteristics along with an associated framework that could be used to describe all of these problems, including debris. The final goal, provided that the first two were possible, was to use the framework to draw comparisons between space debris and the analogous problems. Ultimately, we hoped to provide context and insight for decisionmakers by asking the following question: How have other industries approached their "orbital debris–like" problems? What lessons can be learned from these cases before proceeding with mitigation or remediation measures?[5]

---

[3]   The DARPA Catcher's Mitt study is tasked with the following objectives: model the space debris problem and its future growth; determine which class of satellites is most affected; and, if appropriate, explore technically feasible solutions for debris removal. DARPA intends to use the results of the Catcher's Mitt study to determine whether or not to invest in a space debris remediation program (Jones, undated).

[4]   We define the word *pathfinder* to mean an experimental prototype used to prove a capability.

[5]   Readers seeking additional background on the debris problem are encouraged to obtain a copy of *Artificial Space Debris* by Nicholas L. Johnson and Darren S. McKnight (1987). Alternatively, see Donald J. Kessler and Burton G. Cour-Palais's 1978 paper entitled "Collision Frequency of Artificial Satellites: The Creation of a Debris Belt," which is considered the seminal work on the problem of orbital debris.

This monograph summarizes our methodology, describes the framework that we developed, and highlights the observations that we made when analyzing debris and all of the comparable problems.

CHAPTER TWO

Methodology

We began by examining how perspectives from environmental law, insurance regulation, international relations, policing strategies, and deterrence theory could inform the space debris community from broader, nontechnical perspectives. During discussions with experts with remediation experience from outside the aerospace industry, we examined how technology demonstrations affected the deployment of remediation efforts in these other industries. Furthermore, we investigated key legal, economic, regulatory, and policy concerns that should be considered when evaluating the feasibility of testing technology aimed at reducing orbital debris.

In these discussions, we realized that orbital debris belonged to a group of problems that share a similar set of characteristics. We therefore hypothesized that all of these problems could be evaluated using a single framework, and that we could use this framework to draw comparisons between them.

This idea presented us with the following approach: Use these comparable problems to yield fresh insights on how to think about and deal with the debris problem. For example, how do other industries approach remediation? How has technology development and deployment enabled the remediation efforts? Are there any lessons that could be applied to the debris problem?

We assembled a list of comparable problems that could be analyzed for insights. By choosing to investigate an extensive list of comparable problems, we were able to gather and analyze a set of data from open literature sources that was comprehensive and objective. In addition, this allowed us to ask detailed questions about how these problems evolved over time. For example, how much time elapsed between recognizing that a problem existed and fielding an acceptable solution?

Our primary tool for gathering data was through a literature survey. We also supplemented this review with discussion with experts on several topics that we analyzed. We used the results from these conversations to supplement our understanding and to confirm that our findings were consistent with the established beliefs of the community.

In the end, we gathered the following pieces of information for each of the comparable problems:

1. **Basic overview.** What is the problem?
2. **Calendar dates of key milestones.** When was the problem first identified? When were major mitigation measures imposed? When (if at all) were remedies fielded?
3. **Stakeholder demographics.** Who is viewed as having caused the problem? Who is affected by it? How large is each group? How diverse are their interests?
4. **Current status.** What was the status of each of the problems, as of May 2010? Was it being remedied or simply mitigated?

Once we had this information, we set about describing a standard process that could address space debris and any of the comparable problems. To aid in this process, we also designed a set of tools that allowed us to describe the current status of debris and all of the comparables. Once described using this framework, we started looking for similarities and lessons that could be applied to the debris problem.

The remainder of this document effectuates the methodology described in this chapter. In Chapter Three, we introduce the set of comparable problems that we considered throughout the analysis. We also present a list of characteristics that we developed to describe all of the problems. In Chapter Four, we define the key terms of mitigation and remediation, and in Chapter Five, we describe the framework that we developed. Chapter Six contains an analysis that compares each problem's historical progress as it moved from identification to mitigation or remediation. Chapter Seven reviews the concept of mitigation and provides some effective lessons from other industries. Chapter Eight explores the concept of remediation, and we use the 2010 Deepwater Horizon (DH) oil spill as a case study to draw some conclusions about the nature of an effective remedy. In Chapter Nine, we summarize the important conclusions from the earlier chapters, and we provide some overall observations on the entire analysis.

This document also contains some helpful appendixes. Appendix A contains a brief list of the comparable problems that we identified and considered throughout this project. Appendix B summarizes the current status of each of these problems.

CHAPTER THREE

Comparable Problems and Identifying Characteristics

The good news about orbital debris is that it is not a unique problem. Several industries have faced analogous challenges over the past century and dealt with them successfully.

We identified the following nine comparable examples for use in this analysis: acid rain, airline security, asbestos, chlorofluorocarbons (CFCs), hazardous waste, oil spills, radon, spam, and U.S. border control.[1] We assume that the reader has a general familiarity with each of these topics, and this level of knowledge will be sufficient to understand the concepts presented in this monograph. In addition, Appendix A contains a table that briefly describes each of these issues.

We chose these problems because they all possess the following characteristics:

1. **Behavioral norms (past and/or present) do not address the problem in a satisfactory manner.**[2] In other words, the existing state of affairs does not (and will not) provide an acceptable solution now or in the future. In most industries, there is

---

[1] We chose these nine problems because they represent a diverse set of issues. Other comparable problems certainly exist; this list is not meant to be exhaustive.

[2] We broadly define "behavioral norms" to include individual, commercial, or government conduct. These norms may be based on individual behavior, government regulations, or standard industry practices. Addressing a problem in a satisfactory manner should be considered the same as reducing risk below tolerance levels. The role that risk plays in finding an acceptable solution is discussed in more detail in Chapter Four.

a set of cultural and behavioral norms that govern acceptable behavior. These norms discourage the majority of individuals from engaging in the unwanted behavior, and the results are usually satisfactory. However, for a problem like orbital debris, having a set of normative behaviors does not provide an acceptable solution. For example, most of the international space community agrees that creating additional debris is not acceptable. Yet, debris creation continues to proliferate for a variety of reasons, despite the established belief that debris is damaging to the orbital environment.

2. **The risk of collateral damage is significant.** If a problem is not self-contained, the actions of one party will affect another. Most often, these actions will manifest themselves as inadvertent casualties ("collateral damage") or damages to a third party's property. This threat of collateral damage necessitates an infrastructure that can protect the interests of all stakeholders. For example, if the owner of one satellite creates debris, the resulting fragments could start a chain reaction affecting other entities' satellites and thus their capability, capital investment, or revenue stream.

3. **There will always be an endless supply of "rule-breakers."** Rule-breakers may violate the prevailing behavioral norms intentionally or by accident; their intent does not matter. What does matter is that the supply of rule-breakers is endless. For example, debris has been created intentionally, by exploding lens caps, ASAT tests, or negligent command and control (C2), and by accident, as when two satellites collide on orbit. Even if everyone agreed to stop creating new debris by tethering lens caps and ceasing ASAT use, existing on-orbit satellites may collide with one another and generate a debris cloud. In addition, new space-faring countries may not possess the technical capability or the financial means to effectively follow existing rules and guidelines. In either case, it is reasonable to assume that new debris will continue to proliferate.

4. **The problem will likely never be considered "solved" because the root cause is difficult to eliminate.** There may be several

reasons behind this inability to achieve "solved" status, but the biggest is often that eliminating the root cause is technically challenging or extremely expensive. At the moment, there is no cost-effective way to remove or relocate threatening debris in orbit. In other cases, eliminating the root cause may simply not be an option. For example, the international community could decide to refrain from using the space environment, and debris would no longer be a concern. Obviously, this would be unacceptable to most space-faring corporations and governments, including the United States. In a best-case scenario, the solution will be an asymptotic approach in which the risk is lowered to a level agreed on by all stakeholders. The "solution" will merely minimize collateral damage or effects to a level that is tolerable.

In addition to the comparable problems that we listed, there were several prospective choices that we considered but did not include in the final analysis. For example, we decided to omit global warming. At first glance, global warming seems like an obvious choice: Like orbital debris, it is a global problem that reaches across international boundaries. However, of all the comparable problems that we considered, global warming was the most politically polarizing, and we were not able to find sources on which both sides of the debate could agree. In the end, we concluded that a comparison with global warming would prove more distracting than enlightening for this analysis.

However, by making this decision, we also eliminated a key comparable problem that, like debris, reached across numerous international boundaries and relationships. None of the nine comparables offer such a diverse set of cross-border and relationship considerations.

This is a fair critique, but all of the comparable problems have stakeholders who possess a set of values similar to a group of nation-states. For example, for acid rain, all of the polluters (i.e., individual factories) have the same motivation and self-interest that a group of countries would have. Each wishes to protect its stakeholders, and each brings different cultural values to the discussion. In addition, local and state laws must be reconciled with one another as federal policymakers decide on the best path forward. Finally, the problem of acid rain cannot be solved unless all parties involved agree on a solution. This situation is no different from the debris problem, in which space-faring countries and corporations must work together to develop effective solutions.

CHAPTER FOUR

Nomenclature

Mitigation and Remediation

Since we will use the terms *mitigation* and *remediation* throughout this document, it is critical to define their meanings and distinguish between them before proceeding with the analysis.

Mitigation refers to a class of actions designed to lessen the pain or reduce the severity of something. Standards, rules, and regulations are common examples of mitigating actions: They do not stop unwanted behavior or completely eliminate undesirable outcomes, but they can reduce the frequency or severity of bad events.

Mitigation measures are aimed at preventing a problem from getting worse. Because of this, an effective mitigation strategy needs to be comprehensive, adaptable, and self-correcting.

By contrast, *remediation* aims to reverse events or stop undesired effects. Remediation is often achieved using a technical innovation to reverse undesired outcomes or eliminate undesired risks.[1] For exam-

---

[1]   In this document, we purposely avoid discussing specific remediation technologies that have been proposed for space debris. However, we also recognize that those who are new to the debris problem need to understand what it means to "remediate debris." Therefore, we will mention two exemplar technologies currently under consideration. One proposed approach would use laser radiation to affect a spacecraft's momentum to accelerate the de-orbiting process. An alternative approach would use robots to reposition (or de-orbit) large pieces of debris. We mention these particular methods simply because they are easy to describe within a few sentences; they are not necessarily the most technologically mature nor have they been widely accepted by the space community as viable solutions.

ple, airports use X-ray machines, magnetometers, and microwave body scanners as part of their screening process.

Remedies are often employed in reaction to something, and this has a few implications about their use. First, remedies are targeted reactions designed to address an event that has already occurred. Because remedies should have a targeted purpose, several remediation strategies may be needed to address the overall problem. Finally, remedies are often (but not always) employed after catastrophic events.

For the specific case of space debris, mitigation refers to any action that slows or prevents the future growth of the debris population. Remediation is any action aimed at reducing or eliminating the population of existing space debris so as to avoid future catastrophe.

CHAPTER FIVE

A Framework for Addressing Orbital Debris and the Comparable Problems

Framework

Orbital debris, as well as all of the comparable problems, is best addressed using a series of increasingly aggressive measures designed to discourage the accidental or intentional creation of debris. This chapter outlines a framework that we developed to describe this step-by-step approach.

The framework, shown in Figure 5.1, is represented by a series of concentric rings, where actions become more aggressive as they move toward the center of the diagram. This concentric geometry highlights an important feature of the approach: As the community moves toward the center (and increasingly aggressive deterrents), the size of the debris-generating population decreases with each inward step.

The first step in addressing orbital debris—or any of the comparable problems—is to identify, characterize, and bound the topic in question, as the problem cannot be addressed unless it is first recognized and understood by the community as being an issue of concern.

Once the problem has been identified, characterized, and bounded, the next step is to set normative behaviors that establish acceptable conduct. Most entities will abide by established norms simply because they exist, and this is usually an effective first step toward reducing the number of entities engaging in unwanted behavior.

The United States Pollution Prevention Act of 1990 is a good example of an action that clearly defines some normative behaviors. This act established the following expectations:

Figure 5.1
Framework Stages via Concentric Rings

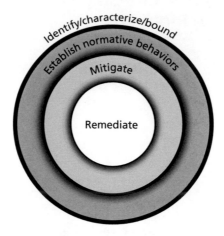

RAND *MG1042-5.1*

Pollution should be prevented or reduced at the source whenever feasible; pollution that cannot be prevented should be recycled in an environmentally safe manner whenever feasible; pollution that cannot be prevented or recycled should be treated in an environmentally safe manner whenever feasible; and disposal or other release into the environment should be employed as a last resort and should be conducted in an environmentally safe manner. (United States Environmental Protection Agency [EPA], 2000)

In the case of orbital debris, the established behavioral norm is that most countries have agreed not to pollute the space environment if they can avoid doing so. This norm is merely a suggestion; there are currently no direct legal or financial penalties for littering in space. However, the U.S. space community has adopted this practice on good faith and does not purposefully release debris into the environment.

Norms tend to discourage unwanted behavior, but some individuals or groups will continue to flout them. To discourage these wrongdoers, the next step is to establish mitigation practices, which may consist of any combination of rules, regulations, standards, incentives, or penalties. These incentive structures are usually very effective at reducing, while not necessarily eliminating, the population of rule-breakers.

Currently, orbital debris mitigation occurs via a number of guidelines that have been established by NASA, the United Nations, and the European Space Agency. For example, in 1993 NASA published a list of guidelines that future U.S. space missions should observe. The behaviors mentioned on the list include the following suggested practices: tethering lens caps, venting spent rocket bodies after separation, and minimizing the use of explosive bolts (NASA, 1993). In practice, U.S. corporations have adopted these measures, and these items are now inspected as part of the final readiness review for every spacecraft launched from the United States. If the spacecraft does not conform to these rules, it is not allowed to launch.

These mitigation practices have been quite effective. Before these measures were adopted, debris caused by exploding spent rocket bodies was the largest contributor to the overall debris population (Prasad, 2005). The rate of incidents for the current generation of spacecraft has been nearly eliminated because of the adoption of these measures.

The final step in addressing the problem is remediation. Normative behavior and mitigation will deter most of the community from engaging in unwanted behavior, but there will always be a handful of rule-breakers. As mentioned in the previous chapter, remediation is a reactionary measure often designed to undo catastrophic damage that has occurred. When a problem is in remediation, the aim is to either relocate the problem's source to a place where it poses less risk or eliminate it entirely. Currently, there are not any cost-effective, operational techniques for remediating orbital debris.

One example from the list of comparable problems that is currently in remediation is airline security. As a result of the 11 September 2001 terrorist attacks, U.S. military jets are now scrambled if a plane is hijacked and there is sufficient evidence to suggest that the plane may be used as a weapon of mass destruction. One remedy, however grim, would be the decision to use an F-16 to shoot down the plane as a last resort. This is a targeted solution to a specific problem, to be used only after terrorists gain control of a plane. This remedy would serve to relocate the problem and minimize the number of innocent deaths by grounding the plane over a region that is less dense with population or industrial activity. As this example demonstrates, remedies can be an important last-ditch contingency option. However, remedies need to be preceded by mitigation, and then tested and proven effective before they can be deployed.

The 2010 DH oil spill is one example where several remedies were ineffective because they had not been tested in real world conditions. The shutoff valves that were designed to close the wellhead in the event of an accident failed to engage. In addition, the oil containment dome—developed in the 1970s to contain spills in shallow waters—initially failed to work because it was not designed for use at a depth of 5,000 feet (Krauss and Saulny, 2010).

Progressing Through the Stages

In practice, the stages outlined in Figure 5.1 are designed for use together to develop an effective way to address the problem. Even as new, more aggressive measures are enacted, decisionmakers should continue to iterate the actions represented by the outer rings, and the lessons gleaned from these iterative attempts can be expected to improve outcomes over time.

The approach must also be flexible, because the problem will inevitably evolve, even as the solution is implemented. New contributing factors to a problem may be discovered, which then require additional mitigation or remedies.

Incorporating the Concept of Risk

The approach shown in Figure 5.1 describes a framework that can be used to address orbital debris as well as each of the comparable problems. However, it does not provide an explanation of how this strategy should be implemented over time. For example, when should decisionmakers conclude that behavioral norms are insufficient and begin implementing mitigation actions? The answer to this question depends on the community's tolerance for risk.[1]

Consider a decisionmaker who oversees a fleet of vehicles charged with transporting a benign chemical substance. The transport vehicles are several years old and occasionally fail, spilling their contents all over the road. The substance is considered to be harmless to living species, so the community is generally tolerant of these spills. The only perceived threat that these spills pose is the resulting traffic jam.

In this example, the community will remain indefinitely just inside the Establish Normative Behaviors ring shown in Figure 5.1. The fleet manager—along with the community—expects the drivers to abide by the rules of the road, but as long as the spills remain occasional and accidental, the community is willing to accept them as a fact of life.

The left-hand plot in Figure 5.2 represents such a scenario. The blue line represents the rate of chemical spills (assumed to be constant through time in this example), and the green line is a notional level of the community's tolerance for these spills. The spills occur less frequently than the public's notional risk threshold, so nothing else will be done to address this problem. In this scenario, normative behaviors are perceived as adequately addressing the problem, and no further actions are taken.

However, suppose that one day a leading scientist presents convincing evidence that contact with the transported chemical may cause immediate death in certain individuals. This development causes the community to change its risk tolerance: People want the frequency of spills reduced by half.

---

[1]   For the purpose of this research, we used Willis's definition of risk, which (paraphrased) is the threat of an unwanted aftereffect (Willis et al., 2005). If the risk eventually turns into reality, the consequences may be realized as (for example) lost revenue, reduced capability, or loss of an asset. Risk can be subdivided into perceived or actuarial risk, both of which can influence a problem. For this research, we will refer to risk only in a general sense, meaning that it can be either perceived or actuarial. We acknowledge that the distinction is important, but specifying this level of detail for each of the analogous problems is outside the scope of this work.

Figure 5.2
Risk Tolerance Versus Undesirable Behaviors (Simple Cases)

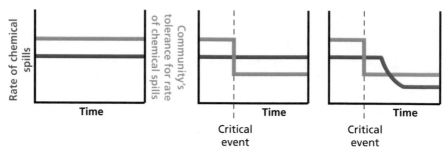

The middle plot in Figure 5.2 represents this scenario. The scientist's announcement (the "critical event") is represented by the step function in the green curve. After the announcement, the community's acceptable level of chemical spills (shown in green) suddenly decreases. The fleet manager is now faced with a crisis.

To address this problem, the decisionmaker needs to proceed to the next ring in Figure 5.1, Mitigate. He sets aside a portion of his budget for vehicle repair, and he authorizes his drivers to use the fund to maintain their vehicles. He also issues a new rule: Any driver who does not maintain his vehicle will be dismissed.

The right-hand plot in Figure 5.2 represents the result of this policy. It takes some time after the scientist's announcement for the decisionmaker to formulate and implement a mitigation policy, but his drivers start maintaining their vehicles, and chemical spills are eventually cut by half. Because the spill rate now falls below the notional tolerance level set by the community, the problem has been adequately addressed using mitigation.

While this example is a simple problem with a straightforward solution, it helps to highlight some important concepts of the framework outlined in Figure 5.1:

- **Decisionmakers must proceed to the next stage when the existing level of unwanted behavior exceeds the community's risk tolerance level.** In this case, mitigation practices were established, and they successfully addressed the problem. However, had these practices only been marginally successful, the decisionmaker would have needed additional mitigation efforts, or he would have needed to proceed to the next ring to pursue a remedy.

- **Eliminating the problem is not necessarily the primary objective.** The primary goal is only to reduce it beneath the community's risk tolerance level. If the decisionmaker tried to eliminate all chemical spills, he might risk bankrupting the company trying to do so. As long as the frequency of spills remains below the tolerance level, the solution is considered adequate, and no additional effort is needed.

- **As mitigation measures are enacted, behavioral norms are still at work.** Even after the decisionmaker institutes a mitigation policy, he still expects his drivers to engage in normative behaviors by abiding by the rules of the road. The result is a compound effect: When implemented together, normative behaviors, mitigation practices, and direct remedies can often result in an effective solution.

Describing Additional Levels of Complexity

The example cited in the previous section is relatively simplistic, but this framework can be adapted to more complex (and realistic) scenarios.

In the hypothetical chemical spill example, neither the spill rate nor the community's tolerance for spills is likely to remain constant over time. This concept is illustrated in the left-hand plot of Figure 5.3. The maximum depicted by the blue line could have been caused by external circumstances, such as a harsh winter that prompted more accidents than usual.

One might assume that the community would become weary of the spills and its tolerance would automatically decrease over time, but this is not necessarily the case. In this example (Figure 5.3, left-hand plot), the tolerance level gradually increases as people realize that death by contact is an exceedingly rare occurrence. In fact, it is also possible

Figure 5.3
Risk Tolerance Versus Undesirable Behaviors (Complex Cases)

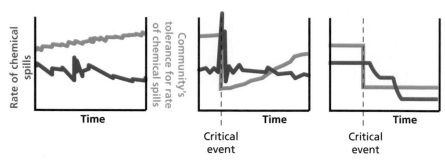

RAND *MG1042-5.3*

that the community's tolerance for these spills will increase over time if the local media simply ceases to report them.

In other circumstances, sudden critical events could have a domino effect: They could simultaneously raise the spill rate and decrease the tolerance threshold. This is shown in the middle plot of Figure 5.3. In a short time after the critical event, the tolerance levels could drop to unreasonably low levels, only to recover after the anomalous event begins to fade from the community's collective memory. The DH is a good example of such an event. In the days immediately after the spill, the U.S. government issued a moratorium on deepwater drilling. However, this act will likely be eased as the country begins to recover and mitigate against future events.

Finally, sometimes it might take several steps of mitigation (or premeditative) efforts before the problem is properly addressed. This concept is illustrated in the right-hand plot of Figure 5.3. When the fleet manager's first response did not sufficiently reduce the problem below the tolerance threshold, he was forced to take additional action: He replaced the oldest vehicles in the fleet with new ones.

CHAPTER SIX

Analysis: Comparing the Timeline of Orbital Debris with the Timelines of the Comparable Problems

In the previous chapter, we used a notional example of chemical spills to describe the conditions under which a fleet manager would move from relying on behavioral norms to implementing mitigation actions. In this chapter, we present a series of timelines that depict how orbital debris and the nine comparable problems have progressed through the stages of identification, establishing behavioral norms, mitigation, and remediation.

As we mentioned at the beginning of Chapter Three, we are using the following nine comparable problems for this analysis: acid rain, airline security, asbestos, CFCs, hazardous waste, oil spills, radon, spam, and U.S. border control. As we mentioned earlier, a general familiarity with these topics is sufficient for understanding the analysis presented in this chapter, but a brief overview of each of these topics is provided in Appendix A. In addition, a detailed summary on the current status of each problem is provided in Appendix B.

Relative Timelines

When evaluating orbital debris and the comparable examples, it is helpful to compare how the problems have evolved in time relative to one another. To do this, we conducted a literature survey on each of the comparable problems. We then determined the length of time

spent in each stage (problem identification, establishment of normative behaviors, mitigation, and remediation) based on research from
periodical sources, legislative records, and court rulings. We also consulted existing analyses from the Congressional Research Service, the
United States Government Accountability Office (GAO), and the
RAND Corporation. The events associated with each problem were
then visualized in a timeline format using a software package designed
for this purpose. Finally, we inspected each timeline and made a judgment about the approximate year in which each problem entered a new
stage. All of the references that we used for this task are listed by topic
in the "Works Consulted for Timelines" section at the end of this document. The result is shown in Figure 6.1, and it provides a notional
comparison that shows how each of the problems progressed through
the four stages.

It is important to note that previous stages do not stop occurring
when a new one begins. This is because these problems continue to
evolve, and the entire strategy—including all of the stages enacted to
the current point—must adapt to these changes. For example, border
control is shown in remediation since the early 1900s, but, even today,
the community continues to reidentify the problem, develop norma-

Figure 6.1
**Notional Comparison of Concentric Ring Progression over Time Across
Orbital Debris and Comparables**

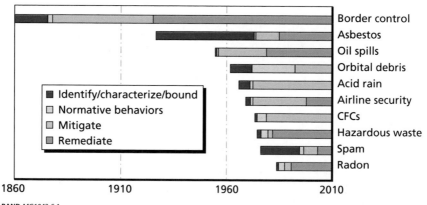

RAND MG1042-6.1

tive behaviors, and implement mitigation strategies to address the latest developments.

The geometry of the bar chart also suggests that a finite boundary exists between each of the stages, but this is generally not the case. For many of the problems, the change from behavioral norms to mitigation or from mitigation to remediation did not happen at a discrete moment in time. Instead, transitions occurred over a series of months or years as legislation or court rulings (for example) were refined or updated.

We used our best judgment to determine where these breaks should occur, based on the list of references that we researched. However, these transitions are approximate and are only meant to provide a graphical representation of the relative progress between the different problems. We did not create this chart with the intent of pinpointing the exact year in which a problem entered a new stage.[1]

The chart shown in Figure 6.1 is useful because it shows the relative lifetime for all of the comparable problems. Border control is—by far—the oldest issue, while spam and radon are newer problems, by comparison. A closer inspection of the chart yields the following observations:

- **It may take several years to identify the problem.** There are several reasons why this may be the case. First, there may be disagreement within the community on whether or not a problem actually exists. In addition, identifying the problem may be difficult because of poor measurement techniques or insufficient communication within the community. When the first asbestos exposure cases were tried in court, there was evidence "that the manufacturers had known about the dangers of asbestos exposure as early as the 1930s" (Carroll et al., 2005). Finally, it may take a while for a phenomenon to reach a critical mass before it is considered a problem. For example, the first piece of spam was sent in May 1978, but spam did not become a serious problem until the Internet linked large numbers of personal computers together.

---

[1]   To emphasize the fact that these transitions are approximate, the horizontal axis has been labeled in 25-year increments instead of in 5- or 10-year increments.

- **In some cases, a single critical event is enough to propel the problem through several stages at once.** These events cause a shock to the status quo and cause the community's risk tolerance to drop precipitously. For example, publicity surrounding the toxic Love Canal incident in the late 1970s caused the American public to suddenly identify hazardous waste as an important public health issue. This concern prompted Congress to pass the Comprehensive Environmental Response, Compensation, and Liability Act (CERCLA) in December 1980 (EPA, undated[a]). Known more commonly as the Superfund, this law empowers the EPA to assign industrial custodians who are then responsible for cleaning up the spill. In this case, hazardous waste went from identification to remediation within a five-year span.
- **A problem does not need a long lifetime in order to enter remediation.** Even if a problem has been around for a relatively short amount of time, it can still enter remediation as quickly as its community deems necessary. For example, once radon in American homes was recognized as a problem, the federal government passed regulations and a massive public awareness campaign ensued to address the problem ("Radon Safety Standard Issued," 1986). The problem entered remediation a mere decade after it was identified.
- **Once in remediation, the problem is not considered "solved."** The comparable problems shown in Figure 6.1 remain important issues, even in the cases in which remediation has been successful. Remediation is often a reaction to a catastrophe, and remedies will always be needed as long as the risk of future catastrophe exists. (Recall that there will always be an endless supply of rule-breakers, as discussed in Chapter Three.) In addition, these problems continue to evolve, and the remediation strategies must evolve to address these new concerns.

CHAPTER SEVEN

Mitigation Strategies and Their Use in Other Communities

This chapter examines the concepts of mitigation in more detail. After reviewing our definition of mitigation, we introduce a set of tools that can be used to identify and describe the stakeholders. This is an important first step because a mitigation strategy will only be effective if it considers the interests and needs of everyone involved.

After describing the stakeholders, we introduce three approaches to mitigation that were originally developed within the context of the environmental protection industry. However, these approaches are general enough that they can be applied to any of the comparable problems, including orbital debris.

Finally, we conclude this chapter with some practical examples of mitigation strategies currently in use. We highlight current mitigation efforts for airline security, radon, and acid rain.

What Is Mitigation?

Mitigation refers to a class of actions designed to lessen the pain or reduce the severity of an outcome. Mitigation does not address the root cause of a problem, but it can effectively treat the symptoms.

Mitigation efforts are usually enacted to reduce the risk associated with a future catastrophe. Because of this, these measures are inherently preventive: They are put in place ahead of time, with the expectation that the measures will treat current and future symptoms. However, in order to be effective, the strategy should be comprehensive enough to address these expected outcomes. Most important, the strategy should also be adaptable to change as the problem evolves.

Describing the Stakeholder Community

When designing a strategy to mitigate against unwanted behavior, the first step is to identify and describe the community that surrounds the problem. This section introduces two metrics that can be used to classify the stakeholder population.

The first tool that we created to describe the stakeholder community uses the metrics of size and diversity. Figure 7.1 shows a notional rendering of the stakeholder space using these metrics.

At one extreme, there is a single stakeholder with a specific set of stated interests. This scenario is often the easiest to mitigate: With one singular entity and set of interests, developing a solution to a problem is usually straightforward.

Figure 7.1
Stakeholder Diversity and Type

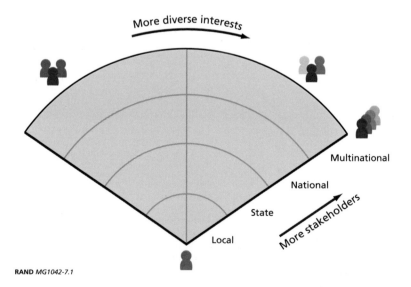

The region in the upper left corner of the fan represents a large community with a homogenous set of interests. Communities that fit these descriptions are usually easy to manage because everyone shares a similar set of values and goals. However, a larger infrastructure may be necessary to manage the larger population.

The upper right portion of the fan represents a large community with a diverse set of interests. Problems that affect these types of communities are the most challenging to mitigate because the mitigation strategy has to satisfy a diverse community with a wide variety of interests.

Another metric that we developed to characterize the population of stakeholders is the overlap between the group that may have caused the problem and the group that is affected by it. This concept is illustrated in Figure 7.2.

Figure 7.2 shows two converging bars, each representing a different population. The top bar is labeled "Blameworthy," and this group represents the community that may be responsible for causing the problem.[1] The lower bar is labeled "Affected," and this group represents the community that is affected by the problem's repercussions.

Figure 7.2
Stakeholder Spectrum: Blameworthy Versus Affected

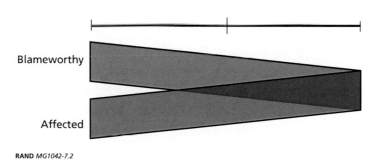

RAND *MG1042-7.2*

---

[1] The choice of the word *blameworthy* is deliberate. We chose this word over *responsible*, *causal*, and *generating* because with many of the problems that we researched, there is some uncertainty within the community over who is responsible for causing the problem. *Blameworthy* suggests a group that is at least associated with creating the problem. Of course, in addition to being at least partially responsible for the problem, the blameworthy party may also be responsible for reconciling the issue.

The intersection of these communities is represented as a spectrum in Figure 7.2. For the purpose of simplifying the explanation, we only considered three discrete points within this spectrum. One extreme of the spectrum is shown on the left side of 7.2. In this extreme, the community is made up of two independent groups: those who generate the problem and those who are affected by the outcome. Suicide bombers on commercial airliners are an example of this extreme. If the bomber is successful, his actions will negatively affect innocent bystanders, and these bystanders have no control or influence over the bomber's decision to commit suicide.

Problems on the left side of the spectrum are very difficult to manage because there is an inherent inequity between the blameworthy and affected parties. Because of this, typical strategies for negotiation based on common goals and compromise tend to fail. For example, a suicide bomber and his intended victims are unlikely to converge on an acceptable compromise.

The other extreme is shown in the right side of Figure 7.2. This extreme represents a complete overlap between those who generate the problem and those who are affected by it. Orbital debris during the 1960s is an example of a problem on this side of the spectrum. During the nascent years of the space age, the United States and the Soviet Union were the only two countries that launched useful payloads into orbit. If one of these countries generated a large debris cloud, it would have affected both nations' space operations.

Problems on the right side of the spectrum can be effectively addressed using mitigation strategies. In the case of the Soviet Union and the United States, there were only two stakeholders, and their interests overlapped. As a result, the problem was easily addressed: Both countries agreed to take care not to intentionally destroy their common operating environment.

Most problems, of course, cannot be categorized at either extreme and instead fall somewhere in the middle, as orbital debris does today. Since the 1960s, many countries—as well as private industry—have developed space capabilities, and that has significantly complicated the task of addressing orbital debris. Not all space-faring nations necessarily share the desire to keep the space environment risk-free. Countries such as Iran or North Korea could be developing abilities to access space with the sole intent of polluting it because this would allow them to counter perceived space-based threats from the United States or another country. If Iran were to purposely create a debris cloud, it would be the blameworthy party, but it would remain relatively unaffected because its society does not have a heavy dependence on space.

The burgeoning commercial space industry represents another community that is pushing the orbital debris problem closer to the middle of the spectrum shown in Figure 7.2. In the past, sovereign nation-states were the only entities with enough resources to field a space capability. These capabilities were developed to benefit their societies economically, scientifically, and socially. These states recognized that space provides a variety of advantages, and they took great care to preserve the environment. They were both the blameworthy and the affected entities.

Today, commercial providers are primarily motivated by profit margins, and they may not have the best interests of their sponsoring country in mind while pursuing their business goals. The commercial provider may be the blameworthy party if it generates debris, but it may not be affected as profoundly by an accident. By contrast, a nation-state that has several hundred active assets on orbit and relies on space-based intelligence to assist ongoing military operations has much more inventory at risk.

Describing the Orbital Debris Community

In the previous section, we introduced the tools illustrated in Figures 7.1 and 7.2 to describe the stakeholder community. We can now use these concepts to summarize the current state of orbital debris. This status is illustrated in Figure 7.3.

The left side of Figure 7.3 shows the debris problem as being in mitigation. Currently, the problem is managed through a series of rules, policies, and regulations that have been established by agencies including the U.S. Air Force, NASA, the European Space Agency, and the United Nations.

Figure 7.3
Orbital Debris: Stakeholder Diversity, Type, and Spectrum

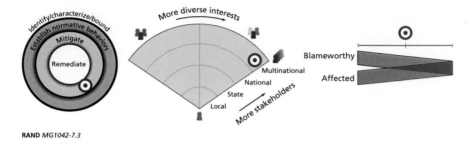

RAND *MG1042-7.3*

We define the stakeholders in this problem as the policymakers and operators for all space-faring nations. This group is multinational and very diverse. Today, private companies such as SpaceX are entering the space lift business based on a for-profit business model. This approach is in contrast to traditional activities, which have always been sponsored by national governments. In addition, smaller countries, such as Iran or North Korea, may plan to use space differently than large countries, such as the United States or Russia, have in the past.

Finally, as we mentioned above, the burgeoning commercial space industry and countries such as Iran and North Korea are pushing the debris problem closer to the middle of the spectrum in Figure 7.2.

Approaches to Mitigation

Several approaches could be used to design a mitigation strategy, but we have chosen to highlight three: C2, market-based, and performance-based approaches. Each of these strategies was developed within the context of the environmental protection industry,[2] but each can also be applied to any of the comparable problems, including orbital debris.

---

[2]   The EPA's *Guidelines for Preparing Economic Analyses* from September 2000 identifies four distinct types of mitigation: traditional design-based C2 approaches, market-oriented approaches, performance-oriented approaches, and nonregulatory approaches. We will discuss the first three strategies below. However, we prefer to categorize the final strategy, non-

We chose these three approaches because they are the most commonly implemented. In addition, at least one of them is currently in use for each of the comparable problems, as we will show below.

Command/Control Approaches

The first approach is commonly referred to as the "command and control" or, more simply, the "C2" approach. In a C2 approach, a community issues a set of commands and then enforces them to exert control over the unwanted behavior, a pattern familiar to Western societies (Raufer and Feldman, 1987). Punishment for transgressors can take the form of monetary damages, but cultural pressures can be an equally effective deterrent. In the case of acid rain, for example, "society's 'commands' are typically given in the form of emission limitations or standards, and the punishment is expressed as fines or, in extreme cases, imprisonment" (Raufer and Feldman, 1987).

The C2 approach is often the first strategy employed against a problem because most societies understand basic incentive structures. Because of this, even when additional mitigation strategies are enacted, a C2 structure is often at work in the background.

A C2 approach can be very effective, particularly on small, singular populations. According to Raufer and Feldman, this approach is also "easy to implement because compliance can be readily determined" (1997). A factory that belches out black plumes of smoke is easily identified as being in violation of emissions rules. Indeed, in its *Guidelines for Preparing Economic Analysis*, the EPA reinforces this sentiment by stating that the "ease of compliance monitoring and enforcement" make C2 strategies an attractive way to mitigate a problem (EPA, 2000).

Even with large, homogeneous populations, a C2 strategy can be effective because one set of regulations is usually sufficient to address the interests of all of the stakeholders. The only change that is required is that a larger enforcement infrastructure is needed with larger populations.

---

regulatory approaches, as part of establishing normative behavior. Therefore, we do not consider this particular strategy to be mitigation using the definition that we outlined earlier.

However, C2 approaches tend to be less effective in controlling large populations with diverse interests because the infrastructure needed to police and enforce the regulations can become too costly and unwieldy. In addition, the limited scope and flexibility of the C2 regulatory structure "may not readily accommodate or encourage technological innovation or may fail to provide incentives to reduce pollution beyond what would be undertaken to comply" with established normative behaviors (EPA, 2000).

Market-Based Approaches

In large or diverse populations, deficiencies or gaps may appear in the C2 system that allow individuals or corporations to engage in the unwanted behavior without repercussions. In such instances, a second mitigation strategy may be required, one that we will loosely call a "market-based" approach. These strategies recognize that some undesirable behavior is acceptable and will occur. A market approach provides an allocation scheme for the right to engage in the undesirable behavior. These strategies are effective at efficiently allocating the right to engage in the behavior, but (by definition) they do a poor job of completely eliminating it.

Market strategies are more challenging to implement than C2 because they require a well-designed, proven infrastructure to administer them (Raufer and Feldman, 1987). Most important, they require 100 percent buy-in from all of the stakeholders, or they will not be effective.

A striking and applicable lesson on the pitfalls of a market approach is offered by Richard Bookstaber in his book, *A Demon of Our Own Design*. Bookstaber's career in the financial markets provided him with the following perspective on market behavior: "In the idealized market, the starting assumption is that the market should run cleanly and transparently. We are faced with more pernicious problems, however, in attaining these goals. When the market ideals collide with the real world, with individuals who are not in control of full information, with institutions that do not act quickly or necessarily in anyone's best interest, the result is like taking a race car for a spin off-road" (Bookstaber, 2007). These words seem particularly wise today, given the financial markets' inability to properly address the collapse of the U.S. housing market in 2008.

Performance-Based Approaches

Performance-based approaches represent yet a third mitigation strategy. These are essentially quota-based initiatives: They prescribe an entity's "maximum allowable level of pollution and then allow the source to meet this target in whatever manner it chooses" (EPA, 2000).

A performance-based approach is useful in the environmental community because the polluting entities can be identified and their emissions are easily measured. While interesting, such an approach is likely not useful to the orbital debris problem because it depends on accurate measurement of the polluting population, and this is a capability that the debris community is currently developing.

Practical Examples of Mitigation

Using the tools summarized above, this section looks at examples of the various mitigation strategies that were applied in practice to the problems of airline security, acid rain, and radon.

Airline Security

Airline security refers to the infrastructure—both at the airport and on the aircraft—that is designed to prevent passengers from threatening commercial air flights while the flight is en route. For this analysis, we only considered security measures for flights that originate or end in the United States.

We define the stakeholders in this problem as the security policymakers, industry workers, and airline passengers; Figure 7.4 illustrates the demographics of this group. We classify the group as mostly homogeneous because nearly all of the stakeholders would agree that passenger safety is paramount (i.e., terrorist acts are unacceptable), and everyone works together to achieve this goal.

Unless the terrorist is a suicide bomber, there is little overlap between the terrorist and those affected. For the sake of completeness,

Figure 7.4
Airline Security: Stakeholder Diversity, Type, and Spectrum

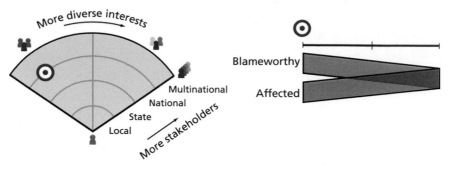

we show this relationship in the right side of Figure 7.4, but it offers little insight for this particular problem. The reason is that the populations are so dissimilar in size: Terrorists represent an extremely small population in comparison to the number of Americans who travel via commercial airliners.

The current commercial airline security policy in the United States is a classic example of a C2 mitigation strategy. As we mentioned above, a C2 approach institutes rules, regulations, and procedures and then enforces them to exert control over a population. In the case of airline security, the Transportation Security Administration (TSA) has established a standard set of procedures that all passengers encounter when they arrive at the airport. For example, passengers are screened using magnetometers and chemical spectrometers, baggage is screened using X-ray machines, and physical security perimeters are established and monitored. Many of these commands are enforced by TSA agents as passengers proceed through the airport and toward the gates.

Given the relatively low incidence of life-threatening violence on commercial air flights, we will assume that this strategy is effective in preventing individuals from terrorizing the passengers on airplanes. One reason this approach is effective is that the stakeholders are mostly homogeneous. Because everyone works toward the common goal of passenger safety, everyone abides by the system of rules, regulations, and procedures.

Another reason why the C2 approach is effective is that the airline environment has discrete physical boundaries. Rules are usually easier to impose and enforce when they are associated with a specific area. The TSA has authority over airport screening areas and limits access to the sterile areas of the airport. This allows the TSA to designate specific areas where their rules, regulations, and procedures are in effect.

While this C2 approach is relatively effective, it is costly. The system requires an enormous infrastructure to handle the thousands of customers passing through each day while monitoring and enforcing all of the rules and regulations. It is only possible because the American public has determined that airline security is a priority and is willing to commit significant tax dollars to mitigate the problem.

The TSA offers a good model for the debris problem because the TSA is entirely focused on mitigation. The C2 approach used in airline security offers some lessons learned that could be applied toward orbital debris:

- **Preparing for potential threats requires an efficient and effective system for collecting and disseminating information.** The infrastructure associated with airline security is always preparing for potential threats. In order to process new intelligence—as well as implement rules—the TSA relies on an effective communication strategy to update officers on the latest procedural changes. In addition, all security-related personnel must have a clearly defined role so that actions can be executed in a prompt manner.
- **An effective mitigation strategy evolves over time.** Throughout the history of airline security, detected threats have led to ongoing revisions in security procedures. The TSA itself was established after the terrorist attacks on 11 September 2001 in order to better mitigate the problem of airline terrorism. New security procedures introduced in response to specific threats have included the use of air marshals, the "3-1-1" rule,[3] and the requirement that passengers remove their shoes for X-ray screening.

---

[3] The so-called "3-1-1" rule dictates that each passenger may carry a 1-quart zip-top bag for the use of carrying liquids in carry-on luggage. Each liquid must be transported in a

- **A successful C2 strategy is enforced by organizations with clearly defined jurisdictions.** In airline security, several different entities may be responsible for maintaining physical security. For example, most airports use a local police force to patrol curbside areas and the tarmac. However, the TSA is responsible for monitoring passengers as they flow through the screening area. If an incident occurs, it is important for all parties to know who is responsible for responding to the alarm.

Radon

Radon is an odorless, tasteless, and radioactive gas that is believed to be the second leading cause of lung cancer in the United States, according to a 2009 assessment by the EPA (EPA, 2009). Radon is a by-product of the naturally occurring decay of uranium, which is present in soils throughout the United States.

The stakeholders in this problem are the policymakers and property owners across the United States; Figure 7.5 illustrates the demographics of this group. We classify the group as mostly homogeneous because most homeowners and nearly all policymakers are interested in minimizing the levels of radon in home and offices.

The mitigation strategy toward radon is rather unique. At first glance, the EPA appears to have implemented a C2 strategy: In 1986, the agency set a standard for radon levels ("Radon Safety Standard Issued," 1986). It recommended that homes containing more than 4 picocuries per liter (pCi/L) be fixed using a number of straightforward remedies.[4] However, the standard was not backed up by any formal incentive structure.

Instead, beginning in the late 1980s, the EPA initiated a massive public awareness campaign designed to promote radon testing by "provid[ing] grants to states to develop programs aimed at encouraging

---

container of 3.4 fl. oz. or less. This policy was instituted in response to a failed terrorist plot designed to detonate liquid explosives on a 2003 transatlantic flight.

[4] According to the EPA, living in a home with 4 pCi/L of radon is equivalent to smoking half a pack of cigarettes a day ("Radon Safety Standard Issued," 1986).

Figure 7.5
Radon: Stakeholder Diversity, Type, and Spectrum

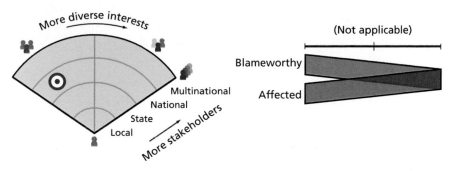

homeowners to test for radon" (GAO, 1992). Today, the results from these efforts are quite evident: A radon test is usually included during the inspection process whenever a home changes ownership.

This mitigation strategy is unique because it relies on the public to voluntarily address the problem. If a home is found to contain levels of radon that are higher than the recommended 4 pCi/L, the EPA does not levy a punishment or a fine. The only penalty that an owner may experience will be a second-order effect: If the owner is unwilling to address the problem, a prospective buyer might be discouraged from purchasing the property.

According to a GAO report, it is difficult to collect accurate data to determine the effectiveness of the public awareness campaign (GAO, 1992). Nevertheless, this type of voluntary approach offers some lessons that could be applied to the orbital debris problem:

- **Nonregulatory approaches may be good at increasing public awareness, but they are unlikely to achieve high levels of compliance.** This type of voluntary approach is similar to one designed to promote seat belt usage in the 1970s. A 1992 GAO report states that a public awareness campaign increased usage from "less than 10 percent [in the mid-1970s] to only 11 percent in 1982" (GAO, 1992). For both seat belts and radon, the initial compliance numbers were very low. This is because the strategy relied on the public to assess the health benefits and determine how they should proceed. This system is certainly very democratic: Every homeowner or car owner is allowed to make an independent decision. However, it is usually very difficult to achieve 100 percent compliance with such a strategy. For radon, this lack of compliance is acceptable because one homeowner's decision not to remove the radon from his or her home is not likely to affect those outside his or her immediate family. However, for a problem like orbital debris—where the actions of one entity can affect everyone else—a nonregulatory approach is unlikely to yield a satisfactory result.

- **Mitigation is relatively straightforward when the problem can be described and measured accurately.** Unlike the case of airline security, where terrorists tend to operate in secrecy, radon levels are easy to measure and the risks associated with high levels are clearly understood. As a result, every homeowner is in a position to assess the risk and decide on a path that is appropriate for his or her needs. Unfortunately, the space community does not enjoy this level of fidelity in its ability to accurately measure the debris population across all orbits and of all sizes. However, improving the measurement capability would increase the available mitigation options.

Acid Rain

Precipitation is "naturally acidic," but the effect is compounded when large quantities of man-made sulfur and nitrogen emissions are introduced into the atmosphere by industrialized societies (Raufer and Feldman, 1987). Acid rain is a problem because it threatens natural ecosystems by lowering the pH level when it falls in rivers, lakes, and oceans. In addition, acid rain is destructive and can shorten the lifetimes of man-made structures such as buildings and bridges.

The stakeholders in this problem are the policymakers, the polluters, and the communities that are affected by acid rain's effects. Figure 7.6 illustrates the demographics of these entities. While acid rain is

Figure 7.6
Acid Rain: Stakeholder Diversity, Type, and Spectrum

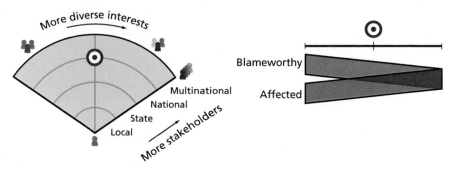

RAND *MG1042-7.6*

geographically concentrated, a problem in the northeast U.S. could affect Canada, so we consider the stakeholders to be multinational.

We also indicate that the stakeholders possess a diverse set of interests. Unlike with airline security and radon, some entities may be resistant to curbing the problem. For example, factories that generate sulfuric emissions but are not harmed by them have balked at the cost of cleanup.

During the 20th century, the United States employed a C2 strategy to mitigate the acid rain problem. The Clean Air Act was first enacted in 1955, and it was subsequently revised in 1970, 1977, and 1990 (McCarthy, 2005). The Clean Air Act set standards, mandates, and deadlines, and it instituted a system of punitive and coercive fines to ensure that entities obeyed the guidelines.

Even today, the EPA continues to issue new regulations to address the current deficiencies in the mitigation plan (Broder, 2010). These recent developments should be interpreted as a positive indicator because, as we pointed out in the discussion about airline security, an effective mitigation strategy must continue to evolve along with the threat.

However, during the latter part of the 20th century, many policymakers began to realize that the existing C2 structure was not addressing the problem in a sufficient manner. To correct for this, Congress passed the Clean Air Act of 1990, which "established an acid rain control program, with a marketable allowance scheme to provide flexibility in implementation" (McCarthy, 2005). As we suggested in the section on market-based approaches, they can be very effective at addressing the deficiencies of a C2 strategy.

Combined, these strategies have been very effective in the United States. For example, in 1990 the federal government "imposed limits on nationwide emissions of sulfur dioxide from power plants and major sources, resulting in a reduction of nearly 50 percent in releases of the chemical" (Broder, 2010). The current market-based approach "seeks to limit, or 'cap,' SO_2 emissions from power plants at 8.95 million tons annually, starting in 2010 [by] authorizing plants to trade SO_2 allowances" (EPA, undated[b]).

The mitigation strategies associated with acid rain in the United States are instructive because acid rain is the only comparable problem that makes use of a market system to reinforce a C2 mitigation strategy. Because of this, there are several lessons that can be applied toward the space debris problem:

- **In order to successfully implement a large C2 strategy, the symptoms must be categorized into groups that represent different levels of relative risk, in categories that range from marginal to moderate to severe.** In the case of the Clean Air Act, each level has a different set of rules, guidelines, incentives, and penalties (McCarthy, 2005). This allows for more efficient deployment of resources, helps to establish jurisdictional boundaries between the different policing entities, and has created a more effective C2 mitigation strategy. For the orbital debris problem, an analogous C2 structure might be categorized by orbit, debris size, or country of ownership.

- **A market approach is often most effective only after an effective C2 strategy is already in place.** A C2 strategy is generally implemented as the first option because it is the most straightforward to design and deploy. However, deficiencies may develop that are best addressed by implementing a market approach to complement the original C2 strategy. In the case of orbital debris, a comprehensive C2 strategy currently does *not* exist, and mitigation resources would likely be best utilized by developing this option first and then implementing a market-based approach at a future date.

CHAPTER EIGHT

Remedies and Their Use in Other Communities

This chapter examines the concepts of remediation in more detail. After reviewing the definition, we describe a distinction that we observed while researching the approaches used to remedy spam, asbestos, and environmental pollution. We also note the important link that often exists between technology and successful remediation. Finally, using the 2010 Deepwater Horizon oil spill as a case study, we highlight the lessons that we compiled after investigating remedies from outside the aerospace industry.

What Is Remediation?

Remediation refers to the act of applying a remedy in order to reverse events or stop undesired effects. Remedies are designed to fix a problem or its effects.

Whereas mitigation is largely a preventive measure, remediation is reactive. Remedies are often applied after the effects of a catastrophe have started to propagate, and they are usually designed to fix a specific effect. Because they can be so specifically targeted, several remediation strategies may be needed to successfully address the broader issue.

Types of Remediation

Remedies can be classified using two sets of descriptive categories, depending on how they address a problem. This section describes these two categories and provides some practical examples taken from the list of comparable problems.

Set 1: Relocation Versus Elimination

An undesired object can be relocated such that it no longer poses a high risk, or it can be completely eliminated. For example, when an oil spill occurs, workers often attempt both relocation and elimination remediation techniques. Skimming techniques are used to remove oil from the ocean's surface so that it may be relocated to a processing plant. In addition, the blameworthy party will plug the source of the spill to eliminate the flow of oil.

The orbital debris problem is unique because either the debris object or the satellite could be relocated to avoid an expected collision. For example, a remedy could be deployed that removes the debris from the satellite's path, or the satellite could avoid the debris by executing a collision avoidance maneuver (Johnson, 2010).

In most cases, elimination is usually a more costly option, and the stakeholder community has to decide which option most appropriately meets its needs. This decision should be based on the community's risk tolerance, as we discussed in Chapter Five.

Set 2: Targeted Versus Dragnet

Undesired objects can be relocated or eliminated using processes that are either targeted or dragnet-like. Targeted removal techniques use a specific method to affect a single, known object. Dragnet strategies indiscriminately trawl space to gather and remove all objects with a particular set of characteristics.

Trawling techniques are most useful in addressing the aftereffects following a catastrophe, such as two satellites colliding or an oil tanker running aground. In both of these examples, a large number of undesired particulates will propagate into the surrounding environment, and a dragnet-type strategy will be the most effective way to collect all of them for relocation or elimination.

Targeted solutions are most effective in addressing the root cause of the problem. For example, once the tanker has run aground, it will continue to leak oil until the rupture is repaired. In this case, a targeted solution is necessary to eliminate the source of problem: A tugboat could be used to dislodge the ship and transport it to an area where the leak can be repaired.

For many problems, it is helpful to be prepared with *both* targeted and dragnet remedies. The fight against email spam is a good example. Most end users remedy spam by applying an email filter (a dragnet), which can be set up to automatically flag messages that are sent from a particular domain name or an unfamiliar sender. This approach is effective in dealing with the problem after the spammer sends the messages.

However, to address the root cause, a targeted approach is necessary, the goal of which is to prevent the spammer from sending the messages in the first place. In this case (and assuming the spammer is operating in the United States), the FBI would learn the spammer's real identify, obtain an arrest warrant, and show up at the person's front door to arrest her or him. Arrest and prosecution will presumably stop the problem at the source.

This example highlights an important feature of the targeted approach: It is often quite effective because the solution is tailored to address a specific problem. However, when a solution is specifically tailored to address a problem, it may lack the flexibility needed to remedy problems that are slightly different.

The Role of Technology

Technology plays an important role in remediation. For the nine comparable problems that we researched for this project, we consider seven of them to be in remediation at some level: airline security, asbestos, hazardous waste, oil spills, radon, spam, and border control. For all of these problems, the remedy is applied through the use of a gadget.[1]

---

[1] We use the word *gadget* simply to mean a physical device or tool that has a useful function. Environmental policy, for example, is not a gadget: By itself, it is not capable of mopping up the latest oil spill or fixing a bridge that has been affected by acid rain. In order to remedy these problems, some sort of physical tool must be deployed.

For many of these problems, the gadget does not have to be state of the art (and therefore expensive) in order to be effective. For example, one common remedy for treating high levels of radon is to install a ventilation system in and around the building's foundation (EPA, 2010). This is a simple solution that uses old technology—fans—to solve a contemporary problem.

Environmental cleanup techniques are often quite simple. For example, cleanup at Superfund sites uses bulldozers to load the contaminated soil into large drums so it can be relocated to a less exposed area.[2] Indeed, little or no technical innovation is needed.

By contrast, some of the comparable problems that we studied could not be remedied without technical innovations. For example, when spam first started to proliferate in the late 1990s, it sparked a market for spam-screening software and eventually prompted developers to offer additional functionality that allowed email clients to automatically apply filters to incoming messages (Giles, 2010).

Oil spills are another example in which state-of-the-art tools are needed for remediation. The operating conditions of deep-sea drilling are quite analogous to those found in space. First, the environment is harsh. The wellhead for deep-sea drilling is often located thousands of feet underwater, and the recovery tools are subject to near-freezing temperatures, extremely high pressure, and corrosive salt water. The tools must be remotely controlled, since humans cannot venture there. In addition, the environment is unpredictable, as Coy and Reed point out in *Bloomberg Businessweek*, "A mile below the surface, things can go to hell in an instant. The pressures and temperatures at work are otherworldly" (Coy and Reed, 2010a).

Similarly, in space, the lack of atmosphere, elevated radiation levels, and extreme temperatures make for a difficult working environment. Things can and do go wrong. Current human space operations are limited to brief extravehicular activities, and this capability will virtually disappear when NASA retires the space shuttle fleet in 2011.

---

[2] The Superfund was established as a way to facilitate cleanup at sites that contain a lot of hazardous waste (EPA, undated[a]). The Superfund provides a means to assess fault, and the blameworthy parties are then held responsible for the cleanup.

It is therefore safe to assume that, for the near term, space operations will be conducted via remote-controlled robots.

These harsh conditions have three implications for the robots that are deployed to fix a problem in space or deep underwater. The first is that these tools must be overengineered. They must have redundant systems in case something fails while the robot is deployed because it will be operated beyond the operator's line of sight. In addition, the robot must be physically hardened such that it can survive a low-temperature, high-pressure environment. Finally, the robot must have an array of sensors so that it may provide adequate situational awareness and feedback to the operator.

These factors all result in a device that will be very expensive to manufacture. In addition, extensive testing will be required to ensure that the robot will be able to operate in the specified environment.

In addition to harsh operating environments, space and ocean exploration are similar in that they are being developed under international ownership treaties. These agreements state that nation-state and corporate entities cannot claim ownership of either environment.

In one respect, orbital debris is actually an easier problem to remedy than oil spills because debris can simply be relocated instead of requiring complete elimination. The space community utilizes a so-called "graveyard orbit," located several hundred kilometers outside the GEO belt, where some aging satellites are relocated before they lose attitude control. This orbit is far enough away as to not interfere with any operational satellites, and they will presumably only cause future conjunction concerns for satellites that are launched from Earth into deep space.[3]

By contrast, oil spill cleanup crews do not have the luxury of declaring part of the ocean a "graveyard." The world community has decided that oil is not acceptable in any part of the ocean, and it must

---

[3] The space community uses the word *conjunction* to mean an alignment (predicted or actual) between two orbiting bodies at the same place and time. To be fair, eventually even this graveyard orbit will become overpopulated, and debris from the resulting conjunctions will likely start to interfere with the GEO belt. The graveyard orbit should not be viewed as a long-term, sustainable solution.

be completely removed using dispersants, skimming techniques, or surface fires (Howell, 2010).

Finally, the catastrophes associated with both space collisions are oil spills are inherently unique. During the Bay of Campeche oil spill in 1979, *New York Times* columnist Gladwin Hill noted that "the watchword among oil-spill experts, and the irritating fact that plagues them, is that 'every spill is different'" (Hill, 1979). This is similar to the debris problem, where the circumstances surrounding every collision will be unique.

All of these similarities suggest that the remediation techniques used to address accidents in the deep-sea drilling community may have some applicability to space-based operations like debris removal. Based on this hypothesis, we decided to take an in-depth look at remediation efforts associated with the oil spill caused by the Deepwater Horizon.

Case Study: Deepwater Horizon (Gulf Of Mexico) Oil Spill

The Deepwater Horizon (DH) was an ultra deepwater, semisubmersible offshore drilling rig contracted to BP by its owner, Transocean. The rig was capable of drilling wells in excess of 35,000 feet while operating in water depths up to 10,000 feet (Transocean Ltd., undated). At the time of the accident, the rig was located in the Gulf of Mexico on the Macondo Prospect. It was operating in 5,000 feet of water on an oil well that was 18,000 feet deep (Corum et al., 2010).

On 20 April 2010, an explosion occurred on the DH, later sinking the platform and causing the largest oil spill in U.S. history ("Gulf of Mexico Oil Spill [2010]," 2010). The DH spill provides a fascinating study in remediation because it highlights several important lessons that we believe are directly applicable to the orbital debris problem.

In the days following the explosion, a number of remedies were deployed in an effort to stop or slow the flow of oil from the wellhead. This section briefly recaps each of the attempts made through the end of May 2010.[4] We will refer to these events as we highlight some lessons learned in the next section.

Shortly after the explosion aboard the DH on 20 April, workers attempted to engage the "blowout preventer," which is a switch designed to remotely trigger a hydraulically operated clamp at the wellhead. When triggered, this clamp is supposed to permanently shut off the flow of oil. When the blowout preventer did not work, workers attempted to use a backup "failsafe switch" to achieve the same result, but this was also unsuccessful (Fountain, 2010a). On 22 April, a second explosion occurred, and the DH sunk (Subcommittee on Energy and Environment Staff, 2010).

On 24 April, remotely operated robots were used in an attempt to manually activate the blowout preventer, which was located at the wellhead 5,000 feet underwater (Robertson and Krauss, 2010). Writing for *Bloomberg Businessweek*, Coy and Reed note that "these valves and shears were the last line of defense, a supposedly impenetrable Maginot Line that made the other fail-safes unnecessary" (Coy and Reed, 2010b). However, the robots were unable to manually activate the valve and seal off the well.

On 2 May, BP began drilling a deepwater relief well (Subcommittee on Energy and Environment Staff, 2010). This measure was designed to relieve the pressure within the main (leaking) well such that it could be permanently capped using mud and concrete. Drilling a relief well is a proven remedy for stopping spills like the one caused by the DH, but it takes several months to implement because the wellhead is so far below the ocean's surface.

On 5 May, BP announced that it had succeeded in shutting off one of the three leaks at the wellhead, but this action had no effect on the rate at which oil continued to escape from the well (Dolnick and Robbins, 2010).

---

[4] The research phase for this (RAND) project concluded on 1 June 2010, so we were only able to track the DH spill through 31 May 2010. While limited in scope, this time span contains sufficient information to support the observations that we present at the end of this section.

On 8 May, BP engineers attempted to lower and place a four-story containment dome over the top of the wellhead. This device is shaped like a giant bell jar, and it is designed to capture the escaping oil so it may be pumped to the surface through a long pipe and collected by a tanker ship. This approach failed because "the dome's opening became clogged with gas hydrates, crystalline particles that form when gas and water combine at low temperature and high pressure" (Wheaton, 2010). While containment domes have been in use for several decades, they had never been tested or used at a depth of 5,000 feet (Wethe, 2010).

On 11 May, engineers lowered a smaller containment dome over the wellhead, thinking that the smaller volume "would capture less seawater and therefore be a lower risk of getting clogged by the gas hydrates" (Krauss and Saulny, 2010). However, this attempt was ultimately abandoned because of an "unspecified reason," according to Interior Secretary Ken Salazar (Revkin, 2010).

On 15 May, BP succeeded in placing an insertion tube—basically an "industrial-sized catheter"—into the riser pipe that was the source of the oil leak (Revkin, 2010). The captured oil was then pumped up to a tanker ship at the surface. However, the riser pipe was 21 inches in diameter, and the insertion tube was only 4 inches wide (Dewan, 2010). According to a *New York Times* report, this method only managed to capture 22,000 gallons for the nine days that the procedure was in use. Using a BP estimate for the rate of oil spilled per day, the insertion tube would have only captured 0.08 percent of the total oil spilled while this remedy was in use (Aigner et al., 2010). BP engineers were quick to note that the procedure was only a stopgap measure until a more permanent solution could be implemented. In fact, the entire process was halted when the "top kill" method was initiated a few days later.

On 17 May, federal officials ordered BP to begin drilling a second relief well (Fountain, 2010b) for use as a backup.

On 20 May, federal officials asked BP to stop using chemical dispersants until BP could identify an alternative that would be less toxic to the environment.[5]

On 26 May, BP started the top kill method, which involved pumping heavy mud into the riser pipe in an attempt to slow down the leak (Kaufman and Krauss, 2010).[6] In addition, so-called "junk shot"—rubber golf balls and shredded tires—was also pumped into the pipe with the hope of slowing down the flow of oil. This attempt was abandoned after three days because the mud was not able to counteract the upward pressure of the oil coming out of the well.

On 31 May, engineers used a remotely controlled "diamond-laced wire saw" to cut off the top of the riser pipe, temporarily increasing the flow of oil (Fountain, 2010b). The first attempts at using the saw failed because the blade reportedly got stuck after coming into contact with the junk shot that was used earlier (Fountain, 2010b).

The saw was eventually freed the following day, the pipe was cut, and a "riser package cap" was lowered onto the top of the newly cut pipe. The cap was expected to funnel some of the leaking oil to a surface ship, but it was not expected to stop all of the oil from entering the ocean.

Figure 8.1 displays all of these remediation attempts on a vertical timeline, and the events are plotted alongside the estimated number of spilled gallons. We used BP's "worst case" estimate for the spill because it was the only metric we could find that provided consistent data across the range of dates between 20 April and 31 May 2010 (Aigner et al., 2010). The chart emphasizes the ineffectiveness of the attempted remedies: None of them were able to affect the spill rate.

---

[5]   Dispersants are detergents designed to break the oil slick into smaller droplets that are easier to disperse via physical skimming or natural consumption via microorganisms (Rosenthal, 2010).

[6]   Once the leak was slowed, BP engineers hoped they would be able to permanently seal the wellhead using concrete.

Figure 8.1
Deepwater Horizon Oil Spill: Timeline of Remediation Attempts and
Estimate of Amount Spilled

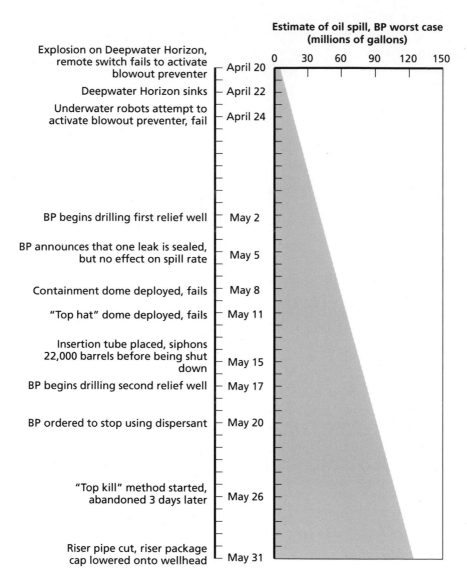

Remediation Lessons Learned from the Deepwater Horizon Spill

The events associated with the DH highlight a number of lessons that are applicable to the orbital debris problem:

- **Simply having a remedy available (or even several) is not sufficient: They must be tested and proven to work in the expected operating conditions.** This is perhaps the biggest lesson of the DH oil spill. The timeline illustrated in Figure 8.1 shows ten discrete attempts to reduce the flow of oil into the Gulf of Mexico, but none were able to affect the spill rate during the first 40 days of the spill. The reason these approaches failed was not because they represented new technology. In fact, a containment dome and the top kill method were also used unsuccessfully in the 1979 Bay of Campeche oil spill off the coast of Mexico (Browne, 1979). Instead, these approaches failed because none of them had been tested to ensure they would work at a depth of 5,000 feet (Wethe, 2010). This concept is applicable to orbital debris because, as we mentioned above, the two environments are very similar. Any future debris removal strategy must be tested to ensure that it will work in the operating environment.
- **The community will only support the development of an effective remedy when the risk posed by the threat is considered to be unacceptable.** We alluded to this concept in Chapter Five, but it is worth mentioning again. In response to the DH spill, Greg Pollock, commissioner of the Oil Spill Prevention and Response Program at the Texas General Land Office said, "We're still dealing with spills the same way we were in the 1960s" (Howell, 2010). For the deep-sea drilling community, the risks associated with a spill have apparently not been significant enough to warrant the development of additional remedies. However, this may change in the months following the DH oil spill. The Gulf States may decide that the risk of a future Deepwater-like accident is not acceptable, institute new laws and regulations aimed at preventing spills, and perhaps force the drilling community to develop

new, more effective remedies. As we demonstrated in Chapter Five, this phenomenon is not unique to oil spills. With radon and asbestos, once the public started to receive information about the negative health effects caused by these substances, they began to demand removal of these substances.

- **When reacting to a catastrophe, a dragnet solution is needed to address the aftereffects.** In the days following the DH spill, a number of workers set out to deploy booms, spray dispersants, and collect animals for cleanup. As we mentioned above, email spam is addressed using filtering software, and Superfund sites are often remedied by simply removing all of the soil from the contaminated area. All of these examples use dragnet-like techniques to identify and remove unwanted pollutants or irritants. This lesson is also applicable to orbital debris. If collisions continue to happen, the space community may decide that the risk posed by the debris population has become unacceptable, and that would likely trigger attempts to reduce the debris population using a dragnet-like technique.

- **After a catastrophe, a targeted solution may also be necessary to remedy a problem.** While dragnet techniques are useful to address the aftereffects of a catastrophe, a targeted technique may be necessary to address the root cause. For the DH spill, the dragnet techniques addressed the resulting oil slick, but they could not be used to stop the leak at the wellhead. To do this, the deep-sea industry needs to develop new techniques (or simply mature existing ones) that effectively shut off wellheads in deep-sea conditions. For orbital debris, the application of a targeted solution is not as obvious. However, if an object were ever launched into space that purposely created orbital debris, a solution would be needed to remedy this problem. In addition, targeted technologies may prove useful in removing or eliminating exceptionally large pieces of debris. However, as large pieces are usually easy to detect and monitor, it might not be cost-effective to spend a significant amount of resources developing a targeted remediation technique that addresses this problem.

- **Remedies must evolve to face the latest challenges.** In the previous section, we mentioned that mitigation approaches must continue to evolve, and the same rule is true for remedies. For example, containment domes were initially developed several decades earlier for use in treating spills at underwater depths of 100–1,000 feet. However, as the oil companies developed new wells at greater depths, this technique did not keep up with the drilling capabilities. As a result, the containment dome approach was not effective when it was deployed to the DH oil spill. This example offers a good lesson for the orbital debris community: Once a remedy is developed, it must be continually monitored and/or improved to ensure that it will be effective under the most current operating conditions.

CHAPTER NINE

Summarizing Observations

This project included three primary objectives: determine whether analogous problems from outside the aerospace industry exist that are comparable to orbital debris; develop a list of identifying characteristics along with an associated framework that could be used to describe all of these problems, including debris; and use the framework to draw comparisons between orbital debris and the analogous problems.

As we mentioned in the introduction, our research identifies effective mitigation strategies, remediation approaches, and lessons learned from outside industries that can be applied to the orbital debris problem.

In this chapter, we summarize the key themes that we identified as we compiled the results from our research. We also provide a detailed look at the question posed at the beginning of introduction: What milestones must be met in order to proceed toward developing mitigation measures or remedial techniques to address orbital debris?

General Observations

We noted the following key themes as we compiled the results from this research:

- **Stakeholders must continuously reassess their situational awareness, risk perception, and ongoing mitigation or remediation efforts.** In order to properly address an issue, the stakeholders need to understand the extent of the problem. An oil spill cannot be remedied until the cleanup crews know where the slick is located. This means that the stakeholders must continuously develop their ability to measure and characterize the problem. For the space community, this means that the stakeholders must have sufficient knowledge of the debris population, or they risk not being able to properly and effectively address the issue.

- **The Superfund could serve as an effective model for orbital debris cleanup.** Assuming that, in the future, the space community decides that the risk posed by debris is too great and it needs to be remedied, the Superfund offers a model for how to accomplish this task. The Superfund was established in 1981, and its goal is to identify the blameworthy parties and facilitate the cleanup process. As Lloyd Dixon notes in a RAND monograph on Superfund liability, "Parties that generated or transported the hazardous material at a site or that owned or operated the site— potentially responsible parties—were held liable for cleaning it up" (Dixon, 2000). This approach represents a rational way to approach the cleanup of orbital debris: It makes the "polluters pay for clean-ups [and] creates strong incentives" for nation-states and private industry to take appropriate preventative measures to avoid creating more debris (Dixon, 2000).

- **Incentive structures (associated with mitigation strategies) work best in the short term. In order to achieve a cost-effective, long-term solution, it is necessary to change stakeholder preferences.** This is an interesting concept culled from the study of deterrence, which is the art of discouraging—or simply intending to discourage—someone from engaging in unwanted activity (Mueller, 2009). When implementing an incentive structure, a command authority is likely to achieve the desired result, but the stakeholder community may not necessary agree with the rules that they are following. In order to achieve effective change, the stakeholders must be convinced that they should adopt these mitigation rules as normative behaviors. For orbital debris, the entire space community needs to agree that purposely creating debris is not acceptable behavior.

- **All of the stages (identify, set normative behaviors, mitigate, and remediate) must continue to evolve with the problem.** The approach used to address a problem must be able to adapt as the problem (and its stakeholders) change over time. For this reason, the current status for any of the stages should never be considered complete. Each community should be constantly reidentifying the problem, standardizing new behavioral norms, refining mitigation techniques, and, if necessary, developing new remedies. Each of these comparable problems resembles a living organism, one that adapts alongside a changing environment. For example, if someone suddenly launches 1,000 microsatellites, the debris problem will require a whole new approach, starting at the outer ring with identify, characterize, and bound.
- **A community cannot enact effective mitigation strategies and/or remedies until the stakeholders agree on an acceptable level of risk tolerance.** As we described in Chapter Five, a community's risk tolerance is an important part of a decisionmaker's calculus. Implementing mitigation and/or remedial measures are only intended to reduce the unwanted behavior to below acceptable risk tolerance levels. Therefore, it is essential that the stakeholders understand and agree on these levels before attempting to address the problem.

The Case for Additional Mitigation

When viewed in light of the comparable problems, there is evidence to suggest that orbital debris does not pose a great enough risk to warrant the deployment of a remediation technology.[1] Currently, the space community appears unwilling to invest in such a venture.

As we suggested in Chapter Five, a community will only move on to the next stage when the risk of the status quo comes to be viewed as

---

[1] The use of the word *deployment* is intentional: It implies an operational—and not simply a pathfinder—system.

unacceptably high. While everyone in the space community certainly agrees that orbital debris poses a risk, the lack of government and private industry funding for this effort suggests that the risk has not yet crossed a critical threshold. Obviously, in the event that a collision with debris destroyed a valuable space asset, the risk calculus would suddenly change in favor of deploying a remedy immediately.

The current lack of private (nongovernment) funding toward debris remedies is particularly telling. Today, the majority ownership of operational space assets (as a percentage of the total operational population) has shifted from government to commercial industry.[2] For this new majority of commercial stakeholders, "the imperative to create shareholder value entails that any investment in a technical system be guided by its value creation potential" (Brathwaite and Saleh, 2009). In other words, if debris was deemed to represent an unacceptable risk to current or future operations, a remedy would already have been developed by the private sector.

One interesting way to quantify the community's risk appears to be currently under way. At a joint NASA/DARPA meeting on orbital debris in December 2009, the Department of Space Technology (SpaceTech) of the Delft University of Technology in the Netherlands distributed a survey with questions specifically designed to gauge the current risk climate.[3] It will be particularly interesting to review the results of this survey because it may serve as an indication of whether the community feels that a remedy is currently necessary.

With these thoughts in mind, we offer the following observations about mitigating the orbital debris problem:

---

[2]   According to the April 2010 UCS Satellite Database, 41 percent of the world's active, operational satellites are solely commercial; 17 percent are solely military; 18 percent are solely government; and the remaining 24 percent are either multiuse or used for research or scientific purposes. While the UCS database represents only an approximate count of the world's total satellite inventory, it is useful in providing a quick estimate to support our claim (UCS, undated).

[3]   Some exemplar questions from this survey: "Is your organisation [sic] generally concerned about space debris? Does the current space debris situation impact your business/operations? In what time frame should this issue be addressed with significant resources? Is your organisation developing a system to remove space debris? Do current regulations and laws adequately support the issue of debris removal?" (SpaceTech, 2009).

- **Mitigation is likely to be more cost-effective than remediation.** As we highlighted in Chapter Four, mitigation measures are inherently preventive; they are enacted with the intent of preventing a problem from getting worse. Therefore, investing in the appropriate mitigation strategy can be a good capital investment because it can significantly reduce the chance that a catastrophic shock to the community will occur. Orbital debris is different from some of the comparable problems because making the leap to a deployable remedy is likely to be quite costly. Investing a fraction of those dollars in mitigation may reduce the risk to a level that is so low that the system will still able to absorb an occasional shock.

- **Consider adopting a common metric for assessing risk posed by space debris.** The mitigation strategies associated with acid rain, asbestos, CFCs, hazardous waste, radon, and spam are all measured using metrics that have been widely accepted by their respective stakeholder communities. The orbital debris community should consider adopting a common metric that measures the risk associated with a particular orbit. For example, the U.S. Air Force uses "fleet aircraft" to monitor the population of a particular aircraft. For every model (e.g., the F-16), additional maintenance and flight logs are kept for a subset of the total inventory. These fleet aircraft then serve as a control group for the rest of the population. When an issue arises, the data from the problem aircraft are compared with that of the control group. This is an effective way to establish a definition for the "average aircraft." Using metrics such as the number of impacts per square meter per year, the risk posed to a subset of "fleet satellites" could be tracked over time.

The Case for Developing Remediation Technology

As we mentioned above, the lack of funding initiatives currently associated with developing a deployable remedy for orbital debris suggests that the community is not yet ready to develop such a capability. However, there are several lessons from our research that suggest it may be wise to develop a pathfinder system in the near term:

- **The community must be prepared for shocks or catastrophic events.** As discussed in Chapter Six, sometimes a single catastrophic event is sufficient to drastically alter stakeholder perceptions. The Chinese ASAT test and the Iridium/Cosmos collision are likely the cause for the increased interest in orbital debris, including this research. In addition, as we highlighted in Chapter Four, remedies are needed to clean up the aftereffects of such event. Developing the pathfinder technology now for such a remedy may prove to be a wise decision because on-orbit collisions are likely to continue to occur in the future.

- **Remedies must be designed and tested to work under the actual operating conditions.** As we discussed in Chapter Nine, this is the biggest lesson from the DH spill. All of the remedies fielded during the first 40 days of the spill were not effective because they had not been tested or proven to work in deep-water drilling conditions. Fielding a demonstration technology will prove useful only if it will provide operators and engineers with relevant information about the technical performance of the actual working conditions. In addition, decisionmakers will gain important data points on realistic values for recharge times, reaction times, and the magazines associated with any potential remediation technology. Ultimately, the pathfinder must strive toward remedying a realistic problem, or the development will risk being considered purely academic and not operationally useful.

- **Our research shows that—for many problems—having just one remedy is not good enough.** As we mention in Chapter Nine, a remedy is often used to respond to an event that has already occurred. As a result, remediation technology is often very specialized, and several different techniques may be necessary to combat the overall problem. There are examples of this throughout all of the comparable problems: Airline security, asbestos, CFCs, oil spills, radon, and spam all use multiple techniques to remedy a problem. Because of this, it may be wise to begin developing a pathfinder system now so that alternative, tangential methods can be developed more quickly in the future.

- **When a problem's effects are not directly observable, a community is likely to underestimate the risk posed by the effects.** Asbestos and radon are two examples in which the effect—cancer—is not physically obvious, and it may not manifest itself for several decades. Under such circumstances, a community may be lulled into a false sense of security because they do not have a daily reminder of the problem's effects. By contrast, the community that surrounds a polluting factory is likely able to see the effects every day. Orbital debris, unfortunately, belongs to the group of problems for which the risk is not readily observable. Because of this, the community may be simply unaware of the severity of the problem. In this case, the technical community should consider implementing an ongoing, metric-based stakeholder awareness program alongside the development of a technical remedy.

A Brief Overview of Orbital Debris and the Comparable Problems

Table A.1 provides a brief description for orbital debris and each of the comparable problems. Throughout this document, we assume that the reader has only a general familiarity with each of these topics. As these problems are discussed in the text, more information is provided only if it is necessary to understand the problem within the context of the analysis.

Table A.1
Overview of Orbital Debris and Comparable Problems

Problem	Description
Acid rain	The presence of air pollution within the atmosphere tends to increase the natural acidity in raindrops. When acid rain falls in a stream or lake, it can lower the pH and threaten plants and animals. In addition, acid rain is destructive to man-made structures, such as buildings and bridges.
Airline security	Commercial air flights represent an attractive target for individuals seeking to terrorize the American public.
Asbestos	Asbestos was used as an insulator and fire retardant in building construction. Extensive medical research suggests a link between asbestos exposure and malignant mesothelioma.
Chlorofluorocarbons (CFCs)	CFCs were used as propellants in consumer aerosol products during the latter part of the 20th century. There is extensive research to suggest that CFCs are responsible for the degradation of the Earth's ozone layer.

Table A.1—Continued

Problem	Description
Hazardous waste	Unauthorized or irresponsible disposal of hazardous waste can threaten local ecosystems and pose a threat to public health. Hazardous waste can include dangerous effluence from an industrial complex; radioactive by-products; or unprocessed human waste that was allowed to enter freshwater streams, rivers, or lakes.
Oil spills	A significant portion of the world's oil is brought to the surface via oil platforms or is transported via tanker ships. When platforms or ships spill oil into streams, lakes, or oceans, the spill represents a threat to environmental ecosystems.
Orbital debris	Objects in LEO (the primary region of concern for most of the space community) traverse at 8 km/s. At these speeds, a collision between two objects—regardless of their size—can be catastrophic.
Radon	Radon is a colorless, odorless, and tasteless gas that is a by-product of uranium decay. It is widespread in the United States, and there is extensive medical research to suggest that exposure causes an increase in lung cancer.
Spam	Spam refers to unsolicited email and/or pop-up ads that may appear when browsing the World Wide Web. The word *spam*—not to be confused with the (trademarked) Spam meat product made by the Hormel Foods Corporation—is likely derived from a Monty Python skit in which the word is used as an unwanted irritation.
U.S. border control	The U.S. Customs Border Patrol was established in 1853 to secure U.S. land borders from unauthorized entrants. Today, access into the United States is tightly controlled via screening areas at all major points of entry by air, land, and sea to address issues such as illegal immigration, drug smuggling, and terrorism.

The Descriptive Framework Applied to Orbital Debris and the Comparables

The following figures and tables provide more detail for orbital debris and each of the comparables.

Figure B.1
Acid Rain

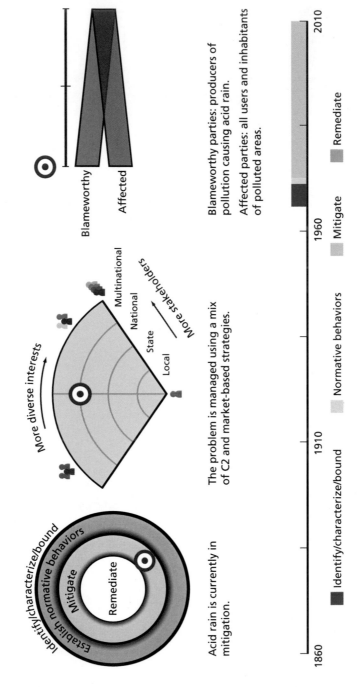

Acid rain is currently in mitigation.

The problem is managed using a mix of C2 and market-based strategies.

Blameworthy parties: producers of pollution causing acid rain.

Affected parties: all users and inhabitants of polluted areas.

Table B.1
Acid Rain

Framework Stage	Description
Identify/characterize/bound problem	Acid rain is a threat to ecosystems and organisms as well as certain man-made structures and manufacturing materials. Acid rain–causing sulfates, acids, and particulates can dwell in the atmosphere and travel long distances before being released via precipitation.
Normative behaviors	Normative behaviors include avoiding emitting harmful sulfates, acids, and particulates into the atmosphere as a by-product of normal operations of a facility or piece of equipment.
Mitigation	The C2 portion of mitigation involves the use of State Implementation Plans, which carry federal enforcement, recordkeeping, and monitoring requirements but saddle states with day-to-day operations. The market-based portion uses a trading program that acts as a mechanism for pollutants to be emitted under certain circumstances and allowances.
Remediation	The acid rain problem is currently not in remediation. One could argue that acid rain is in remediation because the atmosphere will gradually disperse and remove pollutants. However, we do not make this argument.

Figure B.2
Airline Security

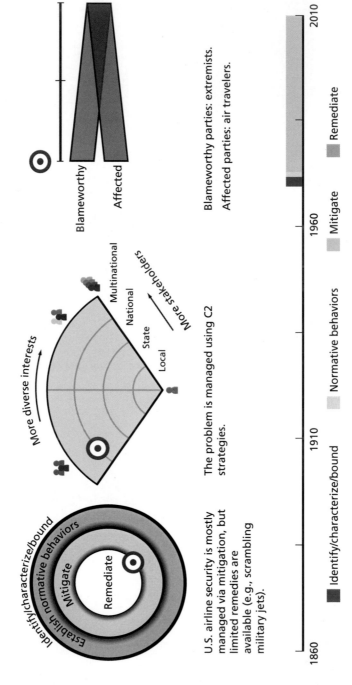

Identify/characterize/bound
Establish normative behaviors
Mitigate
Remediate

U.S. airline security is mostly managed via mitigation, but limited remedies are available (e.g., scrambling military jets).

More diverse interests

Multinational
National
State
Local

More stakeholders

The problem is managed using C2 strategies.

Blameworthy
Affected

Blameworthy parties: extremists.
Affected parties: air travelers.

1860   1910   1960   2010

■ Identify/characterize/bound   ■ Normative behaviors   ■ Mitigate   ■ Remediate

Table B.2
Airline Security

Framework Stage	Description
Identify/characterize/bound problem	Commercial air flights represent an attractive target for individuals seeking to terrorize the American public.
Normative behaviors	American cultural norms suggest that passengers refrain from causing physical or mental distress while on aircraft.
Mitigation	TSA screening procedures reduce the likelihood that dangerous objects or individuals will be present on the flight. Air marshals are prepositioned on certain flights to be available to target and neutralize a passenger who seeks to cause harm when the plane is en route.
Remediation	Military aircraft can be tasked to eliminate a threat if the risk becomes high enough. TSA is responsible for technology demonstrations, although most new technologies fall under the auspices of mitigation.

Figure B.3
Asbestos

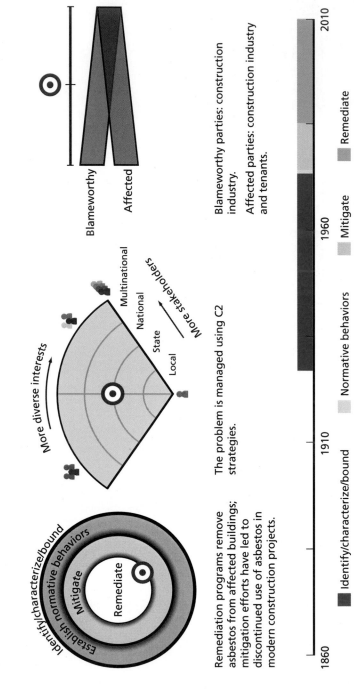

Remediation programs remove asbestos from affected buildings; mitigation efforts have led to discontinued use of asbestos in modern construction projects.

The problem is managed using C2 strategies.

Blameworthy parties: construction industry.

Affected parties: construction industry and tenants.

Identify/characterize/bound

Normative behaviors

Mitigate

Remediate

1860   1910   1960   2010

Table B.3
Asbestos

Framework Stage	Description
Identify/characterize/bound problem	Asbestos is used as an insulator and fire retardant in construction. Over time, a noticeable connection develops between those individuals who have received high exposure to asbestos particles in the air and lung disease and cancer.
Normative behaviors	On determining that a serious health risk existed, asbestos use was mostly discontinued, and alternative materials were identified and substituted.
Mitigation	Rules, regulations, and standards are introduced and exercised to limit exposure and assess conditions for remediation.
Remediation	Processes and procedures are developed and utilized for the controlled removal of asbestos from public buildings in the United States. Technology demonstrations do not play a role in instituting these remedial actions.

Figure B.4
CFCs

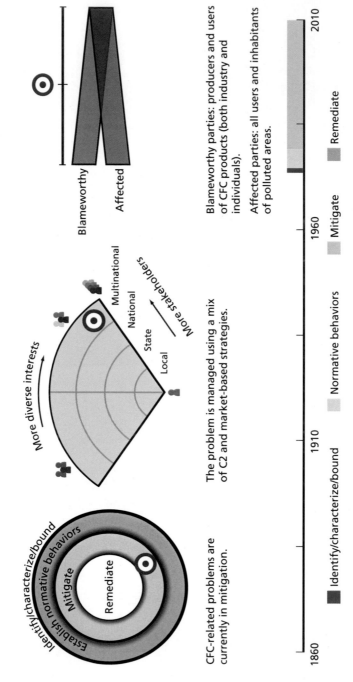

RAND MG1042-B.4

Table B.4
CFCs

Framework Stage	Description
Identify/characterize/bound problem	CFCs, chemical propellants used in consumer aerosol products, were scientifically proven to be responsible for degradation of the Earth's ozone layer.
Normative behaviors	Normative behaviors involved a grass-roots effort by civilian activists to educate the U.S. population regarding the destructive nature of CFCs and thus encourage the population to request formal mitigation.
Mitigation	A C2 mitigation strategy was used to first require explicit labeling of CFC-emitting products and later to ban entirely the sale and use of most products utilizing CFCs.
Remediation	The CFC problem is currently not in remediation. One could argue that the CFC problem is in remediation because the atmosphere will gradually disperse and remove the pollutants. However, we do not make this argument.

Figure B.5
Hazardous Waste

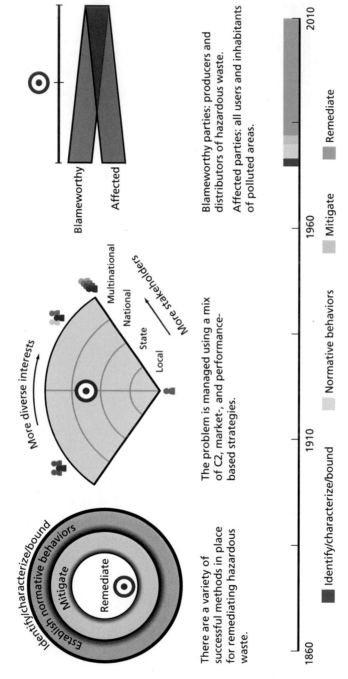

Establish normative behaviors

Identify/characterize/bound

Mitigate

Remediate

More diverse interests

More stakeholders

Local   State   National   Multinational

Blameworthy

Affected

There are a variety of successful methods in place for remediating hazardous waste.

The problem is managed using a mix of C2, market-, and performance-based strategies.

Blameworthy parties: producers and distributors of hazardous waste.

Affected parties: all users and inhabitants of polluted areas.

1860   1910   1960   2010

Identify/characterize/bound   Normative behaviors   Mitigate   Remediate

RAND MG1042-B.5

Table B.5
Hazardous Waste

Framework Stage	Description
Identify/characterize/bound problem	Irresponsible hazardous waste disposal was identified as an action that must be regulated as threats to population health and wellness developed.
Normative behaviors	Known hazardous substances should be disposed of safely, responsibly, and in a controlled manner.
Mitigation	Standards, rules, regulations, and penalties are structured and disseminated. Careful attention is paid to assessment of liability and enforcement of prescribed action.
Remediation	The Superfund is the enabler of remediation. Resources for remediation are allocated based on a stakeholder-developed prioritization scheme. Remediation technologies are relatively rudimentary, effective, but costly to implement.

**Figure B.6
Oil Spills**

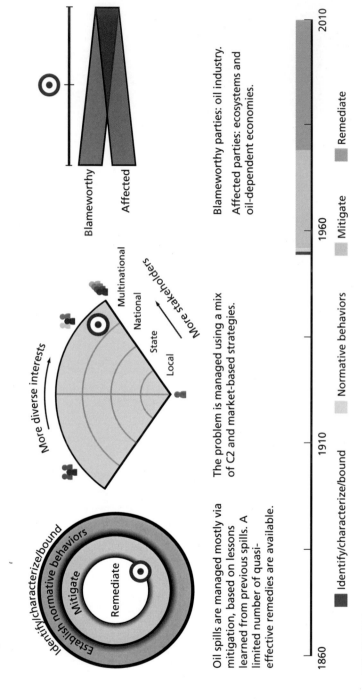

Oil spills are managed mostly via mitigation, based on lessons learned from previous spills. A limited number of quasi-effective remedies are available.

The problem is managed using a mix of C2 and market-based strategies.

Blameworthy parties: oil industry.

Affected parties: ecosystems and oil-dependent economies.

Table B.6
Oil Spills

Framework Stage	Description
Identify/characterize/bound problem	With an increase in the demand for oil, drilling- and shipping-related mishaps resulting in spills also increased. Oil spills are a threat to environmental ecosystems and can involve significant economic losses.
Normative behaviors	Normative behaviors include ensuring regular rig, platform, drill, and ship maintenance as well as responsible operation of these devices.
Mitigation	Laws, standards, and penalties were introduced once the extent of possible environmental impacts exceeded the risk tolerance of the American public. Standards and regulations were also applied to international waters.
Remediation	A myriad of remediation technologies and techniques have been developed and employed with mixed results. Each spill is unique in its characteristics (location, amount of oil, body of water, whether or not the spill source is stationary, etc.), so the remediation toolset must be tailored in response to each event.

Figure B.7
Orbital Debris

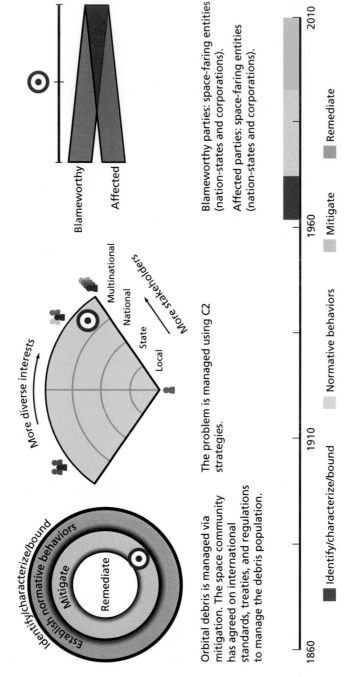

Orbital debris is managed via mitigation. The space community has agreed on international standards, treaties, and regulations to manage the debris population.

The problem is managed using C2 strategies.

Blameworthy parties: space-faring entities (nation-states and corporations).

Affected parties: space-faring entities (nation-states and corporations).

Identify/characterize/bound   Normative behaviors   Mitigate   Remediate

1860   1910   1960   2010

Table B.7
Orbital Debris

Framework Stage	Description
Identify/characterize/bound problem	Space preservation is an issue that affects nation-states, commercial interests, and the international science community, and the existence of orbital debris threatens access to and use of certain orbits around the Earth. Over time, the population of debris has risen.
Normative behaviors	Normative behaviors include establishing a community realization that outer space is an environment that must be protected and respected. Voluntary amendment of operational procedures and manufacturing designs as well as responsible monitoring and control of orbiting assets works to reduce the generation of debris.
Mitigation	A set of international standards exist that were initially developed, adopted, and followed by the United States and then later proposed to the international community via the U.N. Within the U.S. federal government purview, both the Department of Defense and NASA adhere to stringent manufacturing and operational requirements.
Remediation	Orbital debris is not in remediation.

**Figure B.8
Radon**

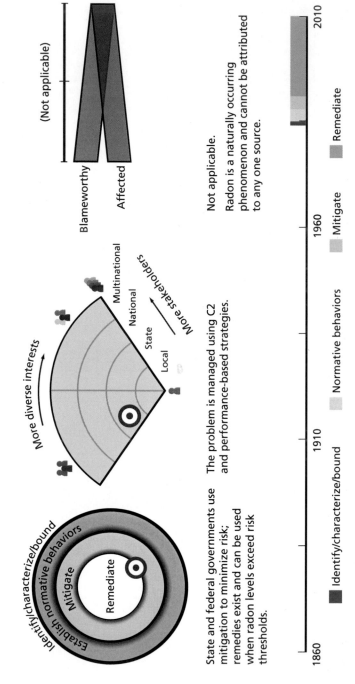

Blameworthy

Affected

(Not applicable)

More diverse interests

Multinational

National

State

Local

More stakeholders

Identify/characterize/bound

Establish normative behaviors

Mitigate

Remediate

State and federal governments use mitigation to minimize risk; remedies exist and can be used when radon levels exceed risk thresholds.

The problem is managed using C2 and performance-based strategies.

Not applicable. Radon is a naturally occurring phenomenon and cannot be attributed to any one source.

1860   1910   1960   2010

Identify/characterize/bound   Normative behaviors   Mitigate   Remediate

RAND MG1042-B.8

Table B.8
Radon

Framework Stage	Description
Identify/characterize/bound problem	Health-threatening levels of naturally occurring radon were found present in homes and other buildings. Multistate surveys were performed in an effort to most accurately estimate how widespread the problem might be.
Normative behaviors	State and local governments began to perform due diligence checks of radon levels in existing structures and at new construction sites.
Mitigation	In lieu of burdensome rules and regulations, mitigation of radon was approached using a public education, nonregulatory approach of suggesting that radon tests be performed in conjunction with real estate transactions.
Remediation	Remediation is recommended when a radon test indicates levels above the tolerance threshold. Remediation techniques are very straightforward and simple and require little to zero advanced technologies.

Figure B.9
Spam

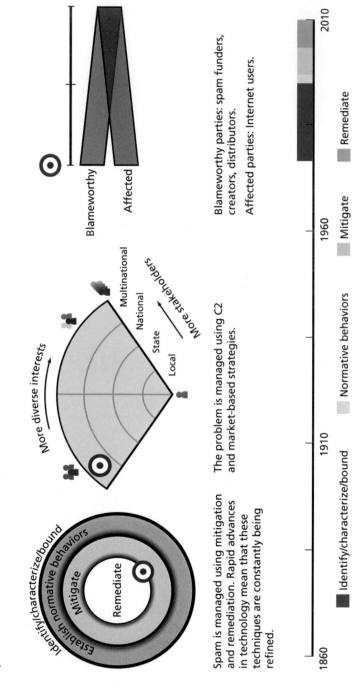

Spam is managed using mitigation and remediation. Rapid advances in technology mean that these techniques are constantly being refined.

The problem is managed using C2 and market-based strategies.

Blameworthy parties: spam funders, creators, distributors.

Affected parties: Internet users.

RAND MG1042-B.9

Table B.9
Spam

Framework Stage	Description
Identify/characterize/bound problem	The size and diversity of the Internet user population increased dramatically over time, which in turn increased the number of potential outlets (or receivers) for spam.
Normative behaviors	Before the adoption of spam techniques by commercial entities because of its low-cost high-distribution characteristics, it was simply considered bad form to "spam" friends, family, colleagues, etc.
Mitigation	Rules, regulations, and penalties were put in place in an effort to severely limit the number of accepted distribution channels and outlets for spam.
Remediation	Pop-up blockers and spam filters are added as features to most interactive, Internet-reliant applications.

Figure B.10
U.S. Border Control

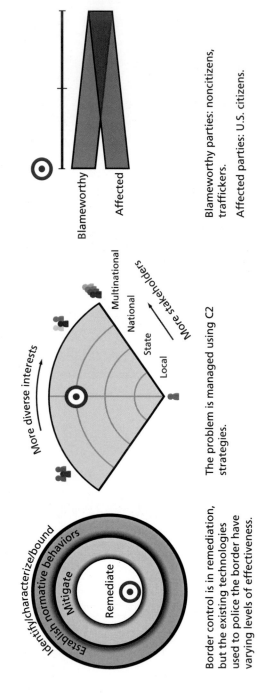

Border control is in remediation, but the existing technologies used to police the border have varying levels of effectiveness.

The problem is managed using C2 strategies.

Blameworthy parties: noncitizens, traffickers.

Affected parties: U.S. citizens.

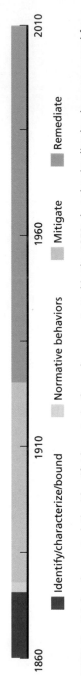

NOTE: For U.S. border control, the identify/characterize/bound process started in the 1850s, but the timeline has been truncated for ease of display.

RAND *MG1042-B.10*

Table B.10
U.S. Border Control

Framework Stage	Description
Identify/characterize/bound problem	U.S. Customs Border Patrol established in 1853 to secure U.S. land borders from unauthorized entrants.
Normative behaviors	The Supreme Court ruled that the federal government is the enforcing body for all things immigration, thus replacing the existing immigration laws passed and enforced at the state level.
Mitigation	National immigration policies, laws, and regulations were formally established and tied to particular agencies and offices for enforcement.
Remediation	Congress adopted an active approach to border control when it established the U.S. Immigration Service Border Patrol, charged with restricting and denying unauthorized access using both passive and active measures. Currently the Secure Border Initiative is tasked with procuring and employing remedial technologies. New technology demonstrations are included in the rollout of remedial capabilities.

Bibliography

Aigner, Erin, et al., "Tracking the Oil Spill in the Gulf," *New York Times*, 2010. As of June 10, 2010:
http://www.nytimes.com/interactive/2010/05/01/us/20100501-oil-spill-tracker.html

Bookstaber, Richard, *A Demon of Our Own Design: Markets, Hedge Funds, and the Perils of Financial Innovation*, Hoboken, N.J.: John Wiley & Sons, 2007.

Brathwaite, Joy, and Joseph H. Saleh, "On Value: Book Value, Market Value, and Hybrid Value Trajectories of Aerospace Companies," paper presented at AIAA SPACE 2009 Conference and Exposition, Pasadena, Calif., September 14–17, 2009.

Broder, John M., "E.P.A. Tightens Rule on Sulfur Dioxide," *New York Times*, June 4, 2010.

Brown, Bernice B., *Delphi Process: A Methodology Used for the Elicitation of Opinions of Experts*, Santa Monica, Calif.: RAND Corporation, P-3925, 1968. As of August 16, 2010:
http://www.rand.org/pubs/papers/P3925/

Browne, Malcolm W., "Mexican Oil Spill Continues, Still Baffling the Experts," *New York Times*, October 5, 1979.

Carrico, Timothy, John Carrico, Lisa Policastri, and Mike Lou, *Investigating Orbital Debris Events Using Numerical Methods With Full Force Model Orbit Propagation*, American Astronautical Society, AAS 08-126, 2008.

Carroll, Stephen J., Deborah R. Hensler, Jennifer Gross, Elizabeth M. Sloss, Matthias Schonlau, Allan Abrahamse, and J. Scott Ashwood, *Asbestos Litigation*, Santa Monica, Calif.: RAND Corporation, MG-162-ICJ, 2005. As of August 16, 2010:
http://www.rand.org/pubs/monographs/MG162/

Corum, Jonathan, et al., "Methods That Have Been Tried to Stop the Leaking Oil," *New York Times*, 2010. As of May 31, 2010: http://www.nytimes.com/interactive/2010/05/25/us/20100525-topkill-diagram. html?ref=us

Coy, Peter, and Stanley Reed, "Oil-Rig Disaster Threatens Future of Offshore Drilling," *Bloomberg Businessweek*, May 14, 2010a. As of June 11, 2010: http://www.bloomberg.com/apps/news?pid=20601109&sid=aHElyJ.bKpsw#

———, "Lessons of the Spill," *Bloomberg Businessweek*, May 6, 2010b. As of May 6, 2010: http://www.businessweek.com/magazine/content/10_20/b4178048176411.htm

Dellapenna, Joseph W., "Law in a Shrinking World: The Interaction of Science and Technology with International Law," Villanova University School of Law Public Law and Legal Theory Research Paper No. 2000-6, June 2000.

Dewan, Shaila, "In First Success, A Tube Captures Some Leaking Oil," *New York Times*, May 17, 2010.

Dixon, Lloyd, *The Financial Implications of Releasing Small Firms and Small-Volume Contributors from Superfund Liability*, Santa Monica, Calif.: RAND Corporation, MR-1171-EPA, 2000. As of August 16, 2010: http://www.rand.org/pubs/monograph_reports/MR1171/

Dolnick, Sam, and Liz Robbins, "BP Says One Oil Leak of Three Is Shut Off," *New York Times*, May 5, 2010. As of June 10, 2010: http://www.nytimes.com/2010/05/06/us/06spill.html

EPA—*See* United States Environmental Protection Agency.

Fiorino, Daniel J., *Making Environmental Policy*, Berkeley, Calif.: University of California Press, 1995.

Fischhoff, B., S. Lichtenstein, P. Slovic, S. L. Derby, and R. L. Keeney, *Acceptable Risk*, New York, N.Y.: Cambridge University Press, 1981.

Fletcher, Susan R., Claudia Copeland, Linda Luther, James E. McCarthy, Mark Reisch, Linda-Jo Schierow, and Mary Tiemann, *Environmental Laws: Summaries of Major Statutes Administered by the Environmental Protection Agency (EPA)*, Congressional Research Service, January 7, 2008.

Fountain, Henry, "Solution to Capping Well Stays Elusive," *New York Times*, April 30, 2010a. As of May 3, 2010: http://www.nytimes.com/2010/05/01/us/01engineering.html

———, "Stuck Saw Delays Effort To Cap Well," *New York Times*, June 3, 2010b.

GAO—*See* United States General Accounting Office.

Giles, Jim, "Spam, Spam and More Spam," *New Scientist*, Vol. 205, No. 2749, 2010, pp. 44–45.

"Gulf of Mexico Oil Spill (2010)," *New York Times*, 2010. As of June 10, 2010:
http://topics.nytimes.com/top/reference/timestopics/subjects/o/oil_spills/gulf_of_
mexico_2010/index.html

Helmer-Hirschberg, Olaf, *Analysis of the Future: The Delphi Method*, Santa
Monica, Calif.: RAND Corporation, P-3558, 1967. As of August 16, 2010:
http://www.rand.org/pubs/papers/P3558/

Hill, Gladwin, "Reporter's Notebook: Oil Slick Mousse," *New York Times*,
August 20, 1979.

Howell, Katie, "Oil Spill Containment, Cleanup Technology Has Failed to Keep
Pace," *Greenwire*, April 30, 2010.

Johnson, Nicholas L., "Orbital Debris: The Growing Threat to Space Operations,"
paper presented at 33rd Annual AAS Guidance and Control Conference,
Breckenridge, Colo., American Astronautical Society, February 6–10, 2010.

Johnson, Nicholas L., and Darren S. McKnight, *Artificial Space Debris*,
Melbourne: Krieger Publishing Company, 1987.

Jones, Johanna, Defense Advanced Research Projects Agency, "Fact Sheet: DARPA
Conducting Study on Orbital Debris Removal," undated. As of June 15, 2010:
http://www.darpa.mil/Docs/OrbitalDebris.pdf

Kaufman, Leslie, and Clifford Krauss, "BP Says Its Latest Effort To Stop Gulf
Leak Failed," *New York Times*, May 30, 2010.

Kehler, General Robert, U.S. Air Force, "Air Force Space Command 'On the
Edge,'" 2010 National Space Symposium, Colorado Springs, Colo., April 13, 2010.

Kessler, Donald J., and Burton G. Cour-Palais, "Collision Frequency of Artificial
Satellites: The Creation of a Debris Belt," *Journal of Geophysical Research*, Vol. 83,
No. A6, 1978, pp. 2637–2646.

Krauss, Clifford, and Susan Saulny, "New Dome Is Prepared To Contain Leaking
Oil," *New York Times*, May 10, 2010.

Matthews, William, "Trackers of Orbiting Junk Sound Warning," *DefenseNews*,
June 10, 2009. As of May 20, 2010:
http://www.defensenews.com/story.php?i=4133281

McCarthy, James E., *Clean Air Act: A Summary of the Act and Its Major
Requirements*, Congressional Research Service, May 9, 2005.

Morgan, M. G., "Probing the Question of Technology-Induced Risk," *IEEE
Spectrum*, Vol. 18, No. 11, 1981a.

Morgan, M. G., "Choosing and Managing Technology-Induced Risk," *IEEE
Spectrum*, Vol. 18, No. 12, 1981b.

Mueller, Karl P., "Don't You Dare: A Very Short Primer on Deterrence," unpublished RAND Corporation research, Santa Monica, Calif., January 28, 2009.

NASA—*See* National Aeronautics and Space Administration.

National Aeronautics and Space Administration, NASA Management Instruction 1700.8: *Policy to Limit Orbital Debris Generation*, April 1993.

———, *Handbook for Limiting Orbital Debris*, NASA-Handbook 8719.14, July 30, 2008.

Orbital Debris Program Office, National Aeronautics and Space Administration, "Orbital Debris: Frequently Asked Questions," undated. As of May 27, 2010: http://orbitaldebris.jsc.nasa.gov/faqs.html#3

Pigou, Arthur C., *The Economics of Welfare*, 4th ed., New York, N.Y.: AMS Press, 1978.

Prasad, M. Y. S., "Technical and Legal Issues Surrounding Space Debris—India's Position in the UN," *Space Policy,* Vol. 21, No. 4, 2005, pp. 243–249.

"Radon Safety Standard Issued," *Facts On File World News Digest*, August 22, 1986.

Raufer, Roger K., and Stephen L. Feldman, *Acid Rain and Emissions Trading: Implementing a Market Approach to Pollution Control*, Totowa, N.J.: Rowman & Littlefield, 1987.

Renn, Ortwin, with annexes by Peter Graham, *Risk Governance: Towards an Integrative Approach*, White Paper #1, Geneva, Switzerland: International Risk Governance Council, September 2005. As of July 16, 2010: http://www.irgc.org/IMG/pdf/IRGC_WP_No_1_Risk_Governance__reprinted_version_.pdf

Revkin, Andrew C., "'Top Hat' Out, Insertion Tube In," *New York Times*, May 14, 2010. As of June 10, 2010: http://dotearth.blogs.nytimes.com/2010/05/14/top-hat-out-insertion-tube-in/

Robertson, Campbell, and Clifford Krauss, "Robots Work to Stop Leak of Oil in Gulf," *New York Times*, April 26, 2010.

Rosenthal, Elisabeth, "In Standoff With Environmental Officials, BP Stays With an Oil Spill Dispersant," *New York Times*, May 25, 2010.

Space Track home page, "Space Track: The Source for Space Surveillance Data," undated. As of June 11, 2010: http://www.space-track.org/perl/login.pl

SpaceTech, Delft University of Technology, NASA-DARPA International Conference on Orbital Debris Removal, Questionnaire on Space Debris Removal, Chantilly, Va., December 8–10, 2009.

Subcommittee on Energy and Environment Staff, House of Representatives Committee on Energy and Commerce, "Memorandum Re: Hearing on 'Combating the BP Oil Spill' on May 27, 2010; Timeline of Events," to members of the Subcommittee on Energy and Environment, Washington, D.C., May 25, 2010.

Transocean Ltd., "Deepwater Horizon Drills World's Deepest Oil & Gas Well," undated. As of June 11, 2010:
http://www.deepwater.com/fw/main/
IDeepwater-Horizon-i-Drills-Worlds-DeepestOil-and-Gas-Well-419C151.html

UCS—*See* Union of Concerned Scientists.

Union of Concerned Scientists, "UCS Satellite Database," undated. As of April 1, 2010:
http://www.ucsusa.org/nuclear_weapons_and_global_security/space_weapons/technical_issues/ucs-satellite-database.html

United States Code, Title 42—The Public Health and Welfare, Chapter 10—Federal Security Agency, January 3, 2007.

United States Environmental Protection Agency, "CERCLA Overview," undated(a). As of May 27, 2010:
http://www.epa.gov/superfund/policy/cercla.htm

———, "Reducing Acid Rain," undated(b). As of March 1, 2010:
http://www.epa.gov/acidrain/reducing/index.html

———, *Guidelines for Preparing Economic Analyses*, Washington, D.C., EPA 240-R-00-003, September 2000.

———, *A Citizen's Guide To Radon: The Guide To Protecting Yourself And Your Family From Radon*, EPA 402/K-09/001, January 2009.

———, *Consumer's Guide To Radon Reduction: How to Fix Your Home*, EPA 402/K-10/002, January 2010.

United States General Accounting Office, *Air Pollution: Actions to Promote Radon Testing*, Washington, D.C., GAO/RCED-93-20, December 1992.

Wethe, David, "BP Oil-Collection Chamber Clogs, Removed From Well (Update1)," *Bloomberg Businessweek*, May 9, 2010. As of September 19, 2010:
http://www.businessweek.com/news/2010-05-09/
bp-oil-collection-chamber-clogs-removed-from-well-update1-.html

Wheaton, Sarah, "Oil Spill Remedies: Trial and Error in Deep Water," *New York Times*, 2010. As of June 11, 2010:
http://www.nytimes.com/interactive/2010/05/28/us/20100528_GULF_TIMELINE.html

Willis, Henry H., Andrew R. Morral, Terrence K. Kelly, and Jamison Jo Medby, *Estimating Terrorism Risk*, Santa Monica, Calif.: RAND Corporation, MG-388-RC, 2005. As of August 16, 2010: http://www.rand.org/pubs/monographs/MG388/

Works Consulted for Timelines

Acid Rain[1]

Raufer, Roger K., and Stephen L. Feldman, *Acid Rain and Emissions Trading: Implementing a Market Approach to Pollution Control*, Totowa, N.J.: Rowman & Littlefield, 1987.

Fowler, David, Rognvald Smith, Jennifer Muller, John Cape, Mark Sutton, Jan Erisman, and Hilde Fagerli, "Long Term Trends in Sulphur and Nitrogen Deposition in Europe and the Cause of Non-Linearities," *Water, Air, & Soil Pollution: Focus*, Vol. 7, No. 1, 2007, pp. 41–47.

United States Environmental Protection Agency, "2008 Highlights: Overview of the Acid Rain Program," 2009. As of March 1, 2010: http://www.epa.gov/airmarkt/progress/ARP_4.html

United States Environmental Protection Agency, Office of Research and Development, National Center for Environmental Research, "Clean Air Research Centers, 2009 Grant Archives, Funding Opportunities," 2009. As of March 2, 2010: http://epa.gov/ncer/rfa/2009/2009_star_clean_air.html

Broder, John M., "E.P.A. Tightens Rule on Sulfur Dioxide," *New York Times*, June 4, 2010.

United States Environmental Protection Agency, "Reducing Acid Rain," undated. As of March 1, 2010: http://www.epa.gov/acidrain/reducing/index.html

---

[1] References in this section were used to build the timeline in Figure 6.1 and are listed in order of publication. Undated references appear at the end of the section.

Airline Security[2]

"Armed Guards for Airlines," *Facts On File World News Digest*, September 16, 1970.

"Antihijacking Program Outlined," *Facts On File World News Digest*, September 23, 1970.

"Nixon Urged to Order Study," *Facts On File World News Digest*, April 15, 1972.

"New Hijacking Curbs Voted," *Facts On File World News Digest*, October 7, 1972.

"Airport Guards to Be Required," *Facts On File World News Digest*, December 2, 1972.

"FAA Tightens Airport Security," *Facts On File World News Digest*, December 25, 1987.

"U.S., Europe Discuss Tougher Steps," *Facts On File World News Digest*, May 12, 1989.

"Study Shows Airport Security Lacking," *Facts On File World News Digest*, December 9, 1999.

Wu, Annie, "History of Airport Security," September 15, 2000. As of May 19, 2010:
http://savvytraveler.publicradio.org/show/features/2000/20000915/security.shtml

"Aviation Security Bill," *Facts On File World News Digest*, November 22, 2001.

"Bush Signs Bill to Federalize Air Security," *Facts On File World News Digest*, November 22, 2001.

"Government Takes Over Airport Security," *Facts On File World News Digest*, February 21, 2002.

United States General Accounting Office, *Progress Since September 11, 2001, and the Challenges Ahead*, Washington, D.C., GAO-03-1150T, September 9, 2003.

Sabasteanski, Anna, "Part 4: Chronology of Significant Terrorist Incidents 1961–2005," in *Patterns of Global Terrorism 1985–2005: U.S. Department of State Reports with Supplementary Documents and Statistics*, Great Barrington, Mass.: Berkshire Pub., 2005.

Public Broadcasting Service, "Timeline: Conflict in the Middle East, 1947–2000," February 14, 2006. As of May 19, 2010:
http://www.pbs.org/wgbh/amex/hijacked/timeline/index.html

---

2   References in this section were used to build the timeline in Figure 6.1 and are listed in order of publication. Undated references appear at the end of the section.

"Airport Security Historical Overview," *Issues & Controversies On File*, June 11, 2007.

Elias, Bart, *Air Cargo Security*, Congressional Research Service, July 30, 2007.

Elias, Bart, *Aviation Security: Background and Policy Options for Screening and Securing Air Cargo*, Congressional Research Service, February 25, 2008.

Elias, Bart, *National Aviation Security Policy, Strategy, and Mode-Specific Plans: Background and Considerations for Congress*, Congressional Research Service, February 2, 2009.

National Counterterrorism Center, *2008 Report on Terrorism*, Office of the Director of National Intelligence, Washington, D.C., April 30, 2009.

United States Government Accountability Office, *A National Strategy and Other Actions Would Strengthen TSA's Efforts to Secure Commercial Airport Perimeters and Access Controls*, Washington, D.C., GAO-09-399, September 2009.

Tibken, Shara, "Airport Security Cos Rise Again As Dutch To Use Body Scanners," *Dow Jones Newswires*, December 30, 2009.

Transportation Security Administration, "311 for Carry-ons," undated. As of June 7, 2010:
http://www.tsa.gov/311/311-carry-ons.shtm

Asbestos[3]

"Asbestos Controls Urged," *Facts On File World News Digest*, October 20, 1971.

"Asbestos Rules Announced," *Facts On File World News Digest*, June 17, 1972.

"EPA Curbs Asbestos, Mercury, Beryllium," *Facts On File World News Digest*, April 21, 1973.

"EPA to Curb 3 Pollutants," *Facts On File World News Digest*, December 22, 1974.

"Occupational Cancer Hazards Reported," *Facts On File World News Digest*, April 19, 1975.

"Asbestos Workers Sue Government," *Facts On File World News Digest*, June 4, 1977.

"EPA Curbed on Asbestos Rule," *Facts On File World News Digest*, January 13, 1978.

---

[3] References in this section were used to build the timeline in Figure 6.1 and are listed in order of publication. Undated references appear at the end of the section.

"Asbestos Workers Get Health Warning," *Facts On File World News Digest*, May 5, 1978.

"Asbestos Rule Revised," *Facts On File World News Digest*, July 7, 1978.

"Asbestos Danger in Schools Cited," *Facts On File World News Digest*, December 31, 1978.

"Asbestos Insurance Cases Refused," *Facts On File World News Digest*, December 18, 1981.

United States General Accounting Office, *Asbestos in Schools: A Dilemma*, Washington, D.C., GAO/CED-82-114, August 31, 1982.

"Tighter Asbestos Rule Proposed," *Facts On File World News Digest*, April 13, 1984.

"Asbestos Removal Rules Revised," *Facts On File World News Digest*, August 30, 1985.

"Asbestos Ban Proposed," *Facts On File World News Digest*, February 14, 1986.

"Asbestos Job Rules Tightened," *Facts On File World News Digest*, June 27, 1986.

"Asbestos Removal Bill Passed," *Facts On File World News Digest*, October 17, 1986.

"Asbestos Removal Planned," *Facts On File World News Digest*, May 1, 1987.

"Schools Urged to Remove Asbestos," *Facts On File World News Digest*, October 30, 1987.

"Asbestos Rules Postponed," *Facts On File World News Digest*, March 18, 1988.

"Asbestos Ban Ruled," *Facts On File World News Digest*, July 14, 1989.

"Asbestos Suits Consolidated," *Facts On File World News Digest*, July 27, 1990.

"Asbestos Alarm Downplayed," *Facts On File World News Digest*, August 8, 1991.

"Court Weakens EPA Asbestos Ban," *Facts On File World News Digest*, October 31, 1991.

United States General Accounting Office, *Asbestos Removal and Disposal: EPA Needs to Improve Compliance With Its Regulations*, Washington, D.C., GAO/RCED-92-83, February 1992.

"Rejection of Asbestos Settlement Upheld," *Facts On File World News Digest*, June 26, 1997.

"Asbestos Risk Seen Exaggerated," *Facts On File World News Digest*, June 11, 1998.

Schierow, Linda-Jo, *The Toxic Substances Control Act: A Summary of the Act and Its Major Requirements*, Congressional Research Service, December 1, 2004.

Carroll, Stephen J., Deborah R. Hensler, Jennifer Gross, Elizabeth M. Sloss, Matthias Schonlau, Allan Abrahamse, and J. Scott Ashwood, *Asbestos Litigation*, Santa Monica, Calif.: RAND Corporation, MG-162-ICJ, 2005. As of August 16, 2010:
http://www.rand.org/pubs/monographs/MG162/

"Senate Shelves Asbestos Bill," *Facts On File World News Digest*, February 16, 2006.

Schierow, Linda-Jo, *The Toxic Substances Control Act (TSCA): Implementation and New Challenges*, Congressional Research Service, July 18, 2008.

CFCs[4]

"Aerosols Said to Pose Ozone Threat," *Facts On File World News Digest*, October 5, 1974.

"Ban on Aerosol Sprays Proposed," *Facts On File World News Digest*, June 21, 1975.

"Aerosol Spray Ban Rejected," *Facts On File World News Digest*, July 19, 1975.

"Aerosol Curb Recommended," *Facts On File World News Digest*, September 18, 1976.

"Aerosol Warning Labels Proposed," *Facts On File World News Digest*, December 4, 1976.

"FDA Requires Fluorocarbon Warnings," *Facts On File World News Digest*, April 30, 1977.

"Fluorocarbon Ban Proposed," *Facts On File World News Digest*, May 21, 1977.

"Fluorocarbon Spray Products Banned," *Facts On File World News Digest*, March 17, 1978.

Palmer, Adele R., William E. Mooz, Timothy H. Quinn, and Kathleen A. Wolf, *Economic Implications of Regulating Chlorofluorocarbon Emissions from Nonaerosol Applications*, Santa Monica, Calif.: RAND Corporation, R-2524-EPA, 1980. As of August 16, 2010:
http://www.rand.org/pubs/reports/R2524/

Wolf, Kathleen A., *Regulating Chlorofluorocarbon Emissions: Effects on Chemical Production*, Santa Monica, Calif.: RAND Corporation, N-1483-EPA, 1980. As of August 16, 2010:
http://www.rand.org/pubs/notes/N1483/

---

[4]   References in this section were used to build the timeline in Figure 6.1 and are listed in order of publication. Undated references appear at the end of the section.

Palmer, Adele R., and Timothy H. Quinn, *Allocating Chlorofluorocarbon Permits: Who Gains, Who Loses, and What Is the Cost?*, Santa Monica, Calif.: RAND Corporation, R-2806-EPA, 1981. As of August 16, 2010: http://www.rand.org/pubs/reports/R2806/

Mooz, W. E., Stephen H. Dole, David L. Jaquette, W. E. Krase, P. F. Morrison, Steven L. Salem, Richard G. Salter, and Kathleen A. Wolf, *Technical Options for Reducing Chlorofluorocarbon Emissions*, Santa Monica, Calif.: RAND Corporation, R-2879-EPA, 1982. As of August 16, 2010: http://www.rand.org/pubs/reports/R2879/

Hammitt, James K., Kathleen A. Wolf, Frank Camm, W. E. Mooz, Timothy H. Quinn, and Anil Bamezai, *Product Uses and Market Trends for Potential Ozone-Depleting Substances, 1985–2000*, Santa Monica, Calif.: RAND Corporation, R-3386-EPA, 1986. As of August 16, 2010: http://www.rand.org/pubs/reports/R3386/

Mooz, W. E., Kathleen A. Wolf, and Frank Camm, *Potential Constraints on Cumulative Global Production of Chlorofluorocarbons*, Santa Monica, Calif.: RAND Corporation, R-3400-EPA, 1986. As of August 16, 2010: http://www.rand.org/pubs/reports/R3400/

"Antarctic Ozone 'Hole' Probed," *Facts On File World News Digest*, August 1, 1986.

"Chemicals Blamed in Polar Ozone Hole," *Facts On File World News Digest*, December 26, 1986.

"Ozone Depletion Feared," *Facts On File World News Digest*, April 17, 1987.

"International Ozone Pact Signed," *Facts On File World News Digest*, September 18, 1987.

"U.S. Ratifies Ozone Treaty," *Facts On File World News Digest*, March 18, 1988.

"CFC Regulation Progresses," *Facts On File World News Digest*, August 5, 1988.

"World Ozone Pact Reached," *Facts On File World News Digest*, July 13, 1990.

"Ozone-Harming Chemicals Declining," *Facts On File World News Digest*, December 19, 1991.

"Ozone Hole Recovery Seen," *Facts On File World News Digest*, December 31, 1999.

United States Senate, "The Clean Air Act (includes amendments through the 108th Congress)," Washington, D.C., 2004.

McCarthy, James E., *Clean Air Act Issues in the 108th Congress*, Congressional Research Service, October 22, 2004.

McCarthy, James E., *Clean Air Act: A Summary of the Act and Its Major Requirements*, Congressional Research Service, May 9, 2005.

United States Code, Title 42, Chapter 85—Air Pollution Prevention and Control, Subchapter I—Programs and Activities, Part A—Air Quality and Emission Limitations, Section 7413—Federal Enforcement, January 3, 2007.

United States Code, Title 42, Chapter 85—Air Pollution Prevention and Control, Subchapter I—Programs and Activities, Part A—Air Quality and Emission Limitations, Section 7414—Recordkeeping, Inspections, Monitoring, and Entry, January 3, 2007.

United States Code, Title 42, Chapter 85—Air Pollution Prevention and Control, Subchapter I—Programs and Activities, Part A—Air Quality and Emission Limitations, Section 7415— International Air Pollution, January 3, 2007.

National Academies press release, "Strong Evidence on Climate Change Underscores Need for Actions to Reduce Emissions And Begin Adapting to Impacts," May 19, 2010.

Environmental Defense Fund, "Clean Air Act Timeline," undated. As of March 26, 2010:
http://www.edf.org/documents/2695_cleanairact.htm

Deepwater Horizon Oil Spill[5]

Pratt, Joseph A., Tyler Priest, and Christopher James Castaneda, *Offshore Pioneers: Brown & Root and the History of Offshore Oil and Gas*, Houston, Tex.: Gulf Publishing Company, 1997.

Melanson, Donald, "Underwater robots to help stem oil spill," *Engadget*, August 24, 2006. As of April 30, 2010:
http://www.engadget.com/2006/08/24/underwater-robots-to-help-stem-oil-spill/

"New Oil Spill Cleanup Technology Developed," *United Press International*, November 14, 2006.

Mascari, Christopher, "OSP Robot Is Human Sized Roomba For Oil Spills," *Gizmodo*, 2008. As of April 30, 2010:
http://gizmodo.com/363718/osp-robot-is-human-sized-roomba-for-oil-spills

United States Environmental Protection Agency, "Oil Pollution Act Overview," last updated March 17, 2009. As of May 1, 2010:
http://www.epa.gov/emergencies/content/lawsregs/opaover.htm

---

[5]   References in this section were used to build the timeline in Figure 8.1 and are listed in order of publication. Undated references appear at the end of the section.

Scripps Institution of Oceanography, "Scripps Scientists to Develop 'Swarms' of Miniature Robotic Ocean Explorers," *Scripps News*, November 10, 2009. As of April 30, 2010:
http://scrippsnews.ucsd.edu/Releases/?releaseID=1031

UCSD Jacobs School of Engineering, "Swarms of Ocean Robots will Drift in Formation, Monitor Oil Spills, Thanks to Advanced Controls Systems," *UCSD Jacobs School of Engineering News*, November 10, 2009. As of April 30, 2010:
http://www.jacobsschool.ucsd.edu/news/news_releases/release.sfe?id=901

Aigner, Erin, et al., "Tracking the Oil Spill in the Gulf," *New York Times*, 2010. As of June 10, 2010:
http://www.nytimes.com/interactive/2010/05/01/us/20100501-oil-spill-tracker.html

Corum, Jonathan, et al., "Methods That Have Been Tried to Stop the Leaking Oil," *New York Times*, 2010. As of May 31, 2010:
http://www.nytimes.com/interactive/2010/05/25/us/20100525-topkill-diagram.html?ref=us

"Gulf of Mexico Oil Spill (2010)," *New York Times*, 2010. As of June 10, 2010:
http://topics.nytimes.com/top/reference/timestopics/subjects/o/oil_spills/gulf_of_mexico_2010/index.html

Wheaton, Sarah, "Oil Spill Remedies: Trial and Error in Deep Water," *New York Times*, 2010. As of June 11, 2010:
http://www.nytimes.com/interactive/2010/05/28/us/20100528_GULF_TIMELINE.html

Moseman, Andrew, "21 Years After Spill, Exxon Valdez Oil Is *Still* Stuck in Alaska's Beaches," *Discover Magazine 80beats*, January 19, 2010. As of April 30, 2010:
http://blogs.discovermagazine.com/80beats/2010/01/19/21-years-after-spill-exxon-valdez-oil-is-still-stuck-in-alaskas-beaches/

Pagnamenta, Robin, and Jacqui Goddard, "Pollution disaster as Deepwater Horizon oil rig sinks into sea," *TimesOnline (London)*, April 23, 2010. As of April 30, 2010:
http://www.timesonline.co.uk/tol/news/environment/article7105649.ece

Goldenberg, Suzanne, "Deepwater Horizon Oil Spill: Underwater Robots Trying to Seal Well," *Guardian*, April 26, 2010. As of April 26, 2010:
http://www.guardian.co.uk/environment/2010/apr/26/deepwater-horizon-spill-underwater-robots

Moseman, Andrew, "Sunken Oil Rig Now Leaking Crude; Robots Head to the Rescue," *Discover Magazine 80beats*, April 26, 2010. As of April 30, 2010:
http://blogs.discovermagazine.com/80beats/2010/04/26/sunken-oil-rig-now-leaking-crude-robots-head-to-the-rescue/

Robertson, Campbell, and Clifford Krauss, "Robots Work to Stop Leak of Oil in Gulf," *New York Times*, April 26, 2010.

Fletcher, Pascal, "Coast Guard Chief Sees Big Risk from Oil Spill," *Reuters*, April 28, 2010. As of April 28, 2010:
http://www.reuters.com/article/idUSTRE63R2ZF20100428

Hadhazy, Adam, "Gulf Oil Spill Is Testing Ground for Future Cleanup Tech," *TechNewsDaily*, April 28, 2010. As of May 1, 2010:
http://www.technewsdaily.com/
gulf-oil-spill-is-testing-ground-for-future-cleanup-tech-0489/

Goldman, Julianna, and Roger Runningen, "BP Will Pay Spill Costs as U.S. Agencies Mobilize (Update3)," *Bloomberg Businessweek*, April 29, 2010. As of April 29, 2010:
http://www.businessweek.com/news/2010-04-29/
bp-will-pay-spill-costs-as-u-s-agencies-mobilize-update3-.html

Krauss, Clifford, "Oil Spill's Blow to BP's Image May Eclipse Costs," *New York Times*, April 29, 2010.

Fountain, Henry, "Solution to Capping Well Stays Elusive," *New York Times*, April 30, 2010. As of May 3, 2010:
http://www.nytimes.com/2010/05/01/us/01engineering.html

Howell, Katie, "Oil Spill Containment, Cleanup Technology Has Failed to Keep Pace," *Greenwire*, April 30, 2010.

Melvin, Jasmin, and Tom Doggett, "Factbox: Major Oil Spills in the United States," *Reuters*, April 30, 2010. As of May 1, 2010:
http://www.reuters.com/article/idUSTRE63T5HZ20100430

Roosevelt, Margot, and Jill Leovy, "Gulf Oil Spill: The Halliburton Connection," *Los Angeles Times*, April 30, 2010. As of May 21, 2010:
http://latimesblogs.latimes.com/greenspace/2010/04/
gulf-oil-spill-the-halliburton-connection.html

Kaufman, Leslie, "New Cleanup Method Holds Hope for Well Leaking About 210,000 Gallons a Day," *New York Times*, May 1, 2010.

Kavanagh, Jim, "Machines and Microbes Will Clean Up Oil," *CNN*, May 1, 2010. As of May 1, 2010:
http://www.cnn.com/2010/US/05/01/oil.spill.geography/index.html

Wald, Matthew L., "Tax on Oil May Help Pay for Cleanup," *New York Times*, May 1, 2010.

Urbina, Ian, Justin Gillis, and Clifford Krauss, "On Defensive, BP Readies Dome to Contain Spill," *New York Times*, May 3, 2010. As of May 4, 2010:
http://www.nytimes.com/2010/05/04/us/04spill.
html?adxnnl=1&adxnnlx=1276016474-P4vuhOl1y7CIIu40DL+WiA

Broder, John M., Campbell Robertson, and Clifford Krauss, "Amount of Spill Could Escalate, Company Admits," *New York Times*, May 4, 2010. As of May 4, 2010:
http://www.nytimes.com/2010/05/05/us/05spill.html

Todd, Brian, "BP to Try Unprecedented Engineering Feat to Stop Oil Spill," *CNN*, May 4, 2010. As of May 4, 2010:
http://www.cnn.com/2010/US/05/03/oil.spill.desperate.measure/index.html

Dolnick, Sam, and Liz Robbins, "BP Says One Oil Leak of Three Is Shut Off," *New York Times*, May 5, 2010. As of June 10, 2010:
http://www.nytimes.com/2010/05/06/us/06spill.html

Resnick-Ault, Jessica, "Detergent-Like Chemicals Turn Oil Into Microbe Snacks (Update1)," *Bloomberg Businessweek*, May 5, 2010. As of September 12, 2010:
http://www.businessweek.com/news/2010-05-05/
detergent-like-chemicals-turn-oil-into-microbe-snacks-update1-.html

Rosenthal, Elisabeth, "In Gulf of Mexico, Chemicals Under Scrutiny," *New York Times*, May 5, 2010. As of May 6, 2010:
http://www.nytimes.com/2010/05/06/science/earth/06dispersants.html

CNN Wire Staff, "'Major Mistakes' in Oil Rig Sinking, Interior Secretary Says," *CNN*, May 6, 2010. As of May 6, 2010:
http://www.cnn.com/2010/US/05/06/gulf.oil.spill/index.html

Coy, Peter, and Stanley Reed, "Lessons of the Spill," *Bloomberg Businessweek*, May 6, 2010. As of May 6, 2010:
http://www.businessweek.com/magazine/content/10_20/b4178048176411.htm

Krauss, Clifford, "For BP, a Battle to Contain Leaks and an Image Fight, Too," *New York Times*, May 6, 2010. As of May 6, 2010:
http://www.nytimes.com/2010/05/07/science/07container.html

Reed, Stanley, "BP's Oil-Spill 'Hive' Buzzes With New Ideas to Stop Leaky Well," *Bloomberg Businessweek*, May 8, 2010. As of May 8, 2010:
http://www.businessweek.com/news/2010-05-08/
bp-s-oil-spill-hive-buzzes-with-new-ideas-to-stop-leaky-well.html

Resnick-Ault, Jessica, and Jim Polson, "Oil-Recovery Box Is BP's Best 'Hope' to Slow Spill (Update1)," *Bloomberg Businessweek*, May 8, 2010. As of May 8, 2010:
http://www.businessweek.com/news/2010-05-08/
oil-recovery-box-is-bp-s-best-hope-to-slow-spill-update1-.html

Robertson, Campbell, "New Setback In Containing Gulf Oil Spill," *New York Times,* May 9, 2010.

Wethe, David, "BP Oil-Collection Chamber Clogs, Removed From Well (Update1)," *Bloomberg Businessweek*, May 9, 2010. As of September 12, 2010:
http://www.businessweek.com/news/2010-05-09/
bp-oil-collection-chamber-clogs-removed-from-well-update1-.html

Krauss, Clifford, and Susan Saulny, "New Dome Is Prepared To Contain Leaking Oil," *New York Times*, May 10, 2010.

CNN Wire Staff, "Next Step to Stop Oil: Throw Garbage at It," *CNN*, May 11, 2010. As of September 12, 2010:
http://www.cnn.com/2010/US/05/09/gulf.oil/index.html

Coy, Peter, and Stanley Reed, "Deepwater Horizon Rig Disaster Threatens Drilling," *Bloomberg Businessweek*, May 14, 2010. As of June 11, 2010:
http://www.bloomberg.com/apps/news?pid=20601109&sid=aHElyJ.bKpsw#

Revkin, Andrew C., "'Top Hat' Out, Insertion Tube In," *New York Times*, May 14, 2010. As of June 10, 2010:
http://dotearth.blogs.nytimes.com/2010/05/14/top-hat-out-insertion-tube-in/

Fountain, Henry, "Throwing Everything, Hoping Some Sticks," *New York Times*, May 15, 2010.

Dewan, Shaila, "In First Success, A Tube Captures Some Leaking Oil," *New York Times*, May 17, 2010.

Ball, Jeffrey, Stephen Power, and Neil King, "U.S. Was Not Ready for Major Oil Spill—Despite Mature Offshore Operations, Gulf Crews Are Improvising With Chemicals, Protective Boom and Outdated Maps," *Wall Street Journal*, May 24, 2010.

Holm, Erik, "Insurance Premiums for Offshore Drilling Soar 15%–50%," *Wall Street Journal*, May 25, 2010.

Rosenthal, Elisabeth, "In Standoff With Environmental Officials, BP Stays with an Oil Spill Dispersant," *New York Times*, May 25, 2010.

Subcommittee on Energy and Environment Staff, House of Representatives Committee on Energy and Commerce, "Memorandum Re: Hearing on 'Combating the BP Oil Spill' on May 27, 2010; Timeline of Events," to members of the Subcommittee on Energy and Environment, Washington, D.C., May 25, 2010.

Krauss, Clifford, "BP Prepares for 'Top Kill' Procedure to Contain Leak," *New York Times*, May 26, 2010.

Power, Stephen, "BP Cites Crucial 'Mistake'," *Wall Street Journal*, May 26, 2010.

Casselman, Ben and Russell Gold, "Unusual Decisions Set Stage for BP Disaster," *Wall Street Journal*, May 27, 2010.

Fountain, Henry, "A Mud That's More Complex Than the Garden Variety," *New York Times*, May 28, 2010.

Kaufman, Leslie, and Clifford Krauss, "BP Says Its Latest Effort To Stop Gulf Leak Failed," *New York Times*, May 30, 2010.

Fountain, Henry, "Stuck Saw Delays Effort To Cap Well," *New York Times*, June 3, 2010.

Fountain, Henry, "Plan for Relief Wells Spurs Hope Amid Caution," *New York Times*, June 3, 2010. As of June 10, 2010: http://www.nytimes.com/2010/06/04/science/earth/04relief.html?pagewanted=1

Leonhardt, David, "Underestimating Risk," *New York Times*, June 6, 2010.

Transocean Ltd., "Deepwater Horizon Drills World's Deepest Oil & Gas Well," undated. As of June 11, 2010: http://www.deepwater.com/fw/main/ IDeepwater-Horizon-i-Drills-Worlds-DeepestOil-and-Gas-Well-419C151.html

Hazardous Waste[6]

"Niagara Site a Disaster Area," *Facts On File World News Digest*, September 15, 1978.

"Love Canal Seepage Known in '58," *Facts On File World News Digest*, April 20, 1979.

"Hazardous Waste Cleanup Proposed," *Facts On File World News Digest*, June 15, 1979.

"Rules for Toxic Wastes Set," *Facts On File World News Digest*, February 29, 1980.

"Toxic Waste Rules Adopted," *Facts On File World News Digest*, May 23, 1980.

"Toxic Waste Clean-Up Bill Cleared," *Facts On File World News Digest*, December 12, 1980.

"Toxic Dump Rules Issued," *Facts On File World News Digest*, July 16, 1982.

"Toxic Waste Regulation Faulted," *Facts On File World News Digest*, March 18, 1983.

"Toxic Waste Bill Signed," *Facts On File World News Digest*, November 16, 1984.

"Superfund Cleanup Guidelines Revised," *Facts On File World News Digest*, February 1, 1985.

"Toxic Cleanup Changes Drafted," *Facts On File World News Digest*, March 8, 1985.

United States General Accounting Office, *EPA's Inventory of Potential Hazardous Waste Sites Is Incomplete*, Washington, D.C., GAO/RCED-85-75, March 26, 1985.

---

[6] References in this section were used to build the timeline in Figure 6.1 and are listed in order of publication. Undated references appear at the end of the section.

United States General Accounting Office, *Cleaning Up Hazardous Wastes: An Overview of Superfund Reauthorization Issues*, Gaithersburg, Md., GAO/RCED-85-69, March 29, 1985.

"Toxic Site Monitoring Faulted," *Facts On File World News Digest*, May 17, 1985.

"House Passes Superfund Bill," *Facts On File World News Digest*, December 13, 1985.

"Reagan Signs Superfund Bill," *Facts On File World News Digest*, October 17, 1986.

United States General Accounting Office, *Hazardous Waste Future Availability of and Need for Treatment Capacity Are Uncertain*, Washington, D.C., GAO/RCED-88-95, April 1988.

"Superfund Cleanup Effort Faulted," *Facts On File World News Digest*, June 24, 1988.

"EPA Report on Hazardous Dump Sites," *Facts On File World News Digest*, November 24, 1988.

"Rules Proposed for Toxic Wastes," *Facts On File World News Digest*, December 22, 1988.

"Superfund Rules Changes Proposed," *Facts On File World News Digest*, April 7, 1994.

"Superfund Reform," *Facts On File World News Digest*, October 13, 1994.

Reisch, Mark, and David M. Bearden, *Superfund Fact Book*, Congressional Research Service, January 27, 1999.

Dixon, Lloyd, *The Financial Implications of Releasing Small Firms and Small-Volume Contributors from Superfund Liability*, Santa Monica, Calif.: RAND Corporation, MR-1171-EPA, 2000. As of August 16, 2010:
http://www.rand.org/pubs/monograph_reports/MR1171/

Esworthy, Robert, *Federal Pollution Control Laws: How Are They Enforced?* Congressional Research Service, February 20, 2008.

United States Environmental Protection Agency, Office of Solid Waste and Emergency Response, *Green Remediation: Incorporating Sustainable Environmental Practices into Remediation of Contaminated Sites*, Washington, D.C., EPA 542-R-08-002, April 2008.

United States Environmental Protection Agency, "CERCLA Overview," undated. As of May 27, 2010:
http://www.epa.gov/superfund/policy/cercla.htm

Oil Spills[7]

"Oil Spill Program," *Facts On File World News Digest*, December 25, 1968.

"Nixon Urges Oil Spill Curbs," *Facts On File World News Digest*, June 3, 1970.

"Strict Oil Spill Rules Set," *Facts On File World News Digest*, July 29, 1970.

"Tough Oil Spill Law Upheld," *Facts On File World News Digest*, April 21, 1973.

"Oil-Spill Legislative Proposals," *Facts On File World News Digest*, July 12, 1975.

Save The River, *The Devastating 1976 NEPCO 140 Oil Spill—Fact Sheet*, 1976.

"Oil-Spill Legislation Requested," *Facts On File World News Digest*, March 26, 1977.

Hill, Gladwin, "Reporter's Notebook: Oil Slick Mousse," *New York Times*, August 20, 1979.

Hovey, Graham, "Liabilities and Alternatives," *New York Times*, August 24, 1979.

"Oil Official Says Mexico Ignored Advice on Well," *New York Times*, September 9, 1979.

Browne, Malcolm W., "Mexican Oil Spill Continues, Still Baffling the Experts," *New York Times*, October 5, 1979.

"Mexican Oil Spill May Continue for Month Despite Cone on Well," *New York Times*, October 17, 1979.

Browne, Malcolm W., "Mexican, Unable to Cap Runaway Oil Well, Plans to Use Chemicals for Cleanup," *New York Times*, October 20, 1979.

Schneider, Eric D., "The Mexican Oil Gushes On and On," *New York Times*, February 26, 1980.

"Mexico Caps Oil Well, Stopping Worst Spill; 3 Million Barrels Lost," *New York Times*, March 25, 1980.

"Ideas & Trends: After 10 Months, The Cap Fits Runaway Ixtoc I," *New York Times*, March 30, 1980.

"Funds Lacking to Complete Studies on Gulf Oil Spill," *New York Times*, April 6, 1980.

"The Lessons of Ixtoc I," *New York Times*, April 12, 1980.

"Oil Spill Reports Required," *Facts On File World News Digest*, July 11, 1980.

---

[7]   References in this section were used to build the timeline in Figure 6.1 and are listed in order of publication. Undated references appear at the end of the section.

"List of World's Largest Accidental Oil Spills," *The Associated Press*, August 7, 1983.

"Birds Dying As Shoreline Clean Up Continues," *The Associated Press*, October 9, 1985.

Foster, David, "Oil Spills Still Fact Of Life After 20 Years Of Environmental Progress," *The Associated Press*, January 14, 1989.

"Largest U.S. Oil Spill Fouls Alaska Marine Habitat," *Facts On File World News Digest*, March 31, 1989.

"Ashland Oil to Pay $4.6 Million to Pennsylvania," *New York Times*, November 23, 1989.

Haddad, Annette, "Harbors, Portion of Beach Reopened," *United Press International*, February 19, 1990.

"Mega Borg Sold for Scrap Iron," *United Press International*, July 17, 1990.

"Oil-Spill Bill Cleared," *Facts On File World News Digest*, August 17, 1990.

"Worst Oil Spills," *The Associated Press*, April 14, 1991.

"List of Worst U.S. Oil Spills," *The Associated Press*, January 7, 1994.

"Rhode Island Oil Spill Is More Serious Than Initially Thought," *New York Times*, January 22, 1996.

"Industry Unveils Anti-Spill Plan," *Facts On File World News Digest*, September 21, 1999.

Larry, M. Lanning, and M. Ford Roseanne, "Glass Micromodel Study of Bacterial Dispersion in Spatially Periodic Porous Networks," *Biotechnology and Bioengineering*, Vol. 78, No. 5, 2002, pp. 556–566.

Alaska Department of Environmental Conservation, Division of Spill Prevention and Response, Prevention and Emergency Response Program, *Joint After-Action Report for the TAPS Bullet Hole Response (October 2001)*, February 8, 2002.

Gray, Geoffrey, "Fishy Doings Surface in Polluted Newtown Creek," *New York Sun*, July 24, 2004.

Graham, Troy, "Oil Spill Is Rated 2d Worst on River; Officials Estimate the Ship Lost 265,000 Gallons. The Probe and Cleanup Should Take Months," *Philadelphia Inquirer*, January 15, 2005.

Berman, Russell, "Greenpoint, Maspeth Residents Lobby To Get 55-Year-Old Oil Spill Cleaned Up," *New York Sun*, November 18, 2005.

Arnold, Scott, *M/V Selendang Ayu Oil Spill, Unalaska, Alaska, Public Health Evaluation of Subsistence, Resources Collected During 2005, Final Report*, April 18, 2006.

Greg Fruge, Jr., *Oil Spill Incident Report—Citgo Oil Spill*, Louisiana Department of Environmental Quality, June 25, 2006.

Ramseur, Jonathan L., *Oil Spills in U.S. Coastal Waters: Background, Governance, and Issues for Congress*, Congressional Research Service, October 26, 2006.

"How Do Oil Spills Impact Casco Bay?" in *Toxic Pollution in Casco Bay: Sources and Impacts*, Portland, Me.: Casco Bay Estuary Partnership, 2007.

United States Code, Title 33—Navigation and Navigable Waters, Chapter 40—Oil Pollution, Subchapter I—Oil Pollution Liability and Compensation, Section 2702—Elements of Liability, January 3, 2007.

United States Code, Title 33—Navigation and Navigable Waters, Chapter 40—Oil Pollution, Subchapter I—Oil Pollution Liability and Compensation, Section 2711—Consultation on Removal Actions, January 3, 2007.

United States Code, Title 33—Navigation and Navigable Waters, Chapter 40—Oil Pollution, Subchapter I—Oil Pollution Liability and Compensation, Section 2716—Financial Responsibility, January 3, 2007.

United States Code, Title 33—Navigation and Navigable Waters, Chapter 40—Oil Pollution, Subchapter I—Oil Pollution Liability and Compensation, January 3, 2007.

United States Government Accountability Office, *Maritime Transportation: Major Oil Spills Occur Infrequently, but Risks to the Federal Oil Spill Fund Remain*, Washington, DC, GAO-07-1085, September 2007.

Curiel, Jonathan, Jane Kay, and Kevin Fagan, "Spill Closes Bay Beaches as Oil Spreads, Kills Wildlife," *San Francisco Chronicle*, November 9, 2007. As of May 14, 2010:
http://www.sfgate.com/cgi-bin/article.cgi?f=/c/a/2007/11/08/BAD8T8PLU.DTL

Alaska Department of Environmental Conservation, Division of Spill Prevention and Response, Prevention and Emergency Response Program, *Situation Report*, March 28, 2008.

United States Fish and Wildlife Service, Environmental Contaminants Program, "New Orleans, Louisiana Oil Spill (Tanker *Tintomara* Collision with Barge DM932)," last updated September 14, 2009. As of May 19, 2010:
http://www.fws.gov/contaminants/examples/BargeLAResponse.htm

Strategic Environmental Research and Development Program (SERDP) and SERDP Exploratory Development (SEED), *In Situ Remediation of Contaminated Aquatic Sediments*, Arlington, VA, ERSEED-11-01, October 29, 2009.

Bureau of Ocean Energy Management, Regulation, and Enforcement, U.S. Department of the Interior, "Oil Spill Response Research (OSRR) Program," 2010. As of May 1, 2010:
http://www.mms.gov/taroilspills/

LiveScience, "LiveScience Topics: Oil Spill," 2010. As of June 7, 2010:
http://www.livescience.com/topic/oil-spill

Gonzalez, Angel, and Naureen Malik, "Collision Causes Crude Oil Spill in Texas," *Wall Street Journal*, January 24, 2010.

Buzzards Bay National Estuary Program, "Overview of the Bouchard 120 Oil Spill," undated. As of May 14, 2010:
http://www.buzzardsbay.org/oilspill-4-28-03.htm

International Tanker Owners Pollution Federation, "Case Histories—A," undated. As of May 18, 2010:
http://www.itopf.com/information-services/data-and-statistics/case-histories/alist.html#argo

International Tanker Owners Pollution Federation, "Statistics—Major Oil Spills," undated. As of May 18, 2010:
http://www.itopf.com/information-services/data-and-statistics/statistics/#major

Orbital Debris[8]

United Nations, *Charter of the United Nations, Chapter 7: Action with Respect to Threats to the Peace, Breaches of the Peace, and Acts of Aggression*, 1945.

Kessler, Donald J., and Burton G. Cour-Palais, "Collision Frequency of Artificial Satellites: The Creation of a Debris Belt," *Journal of Geophysical Research*, Vol. 83, No. A6, 1978, pp. 2637–2646.

Johnson, Nicholas L., and Darren S. McKnight, *Artificial Space Debris*, Melbourne: Krieger Publishing Company, 1987.

Remillard, Stephen K., *Debris Production in Hypervelocity Impact ASAT Engagements*, Wright-Patterson Air Force Base, Ohio: Air University, Air Force Institute of Technology, 1990.

United States Congress, Office of Technology Assessment, *Orbiting Debris: A Space Environmental Problem—Background Paper*, Washington, D.C., OTA-BP-ISC-72, September 1990.

Reibel, David Enrico, "Environmental Regulation of Space Activity: The Case of Orbital Debris," *Stanford Environmental Law Journal*, Vol. 10, 1991, pp. 97–136.

Hunter, Major Roger C., *A US ASAT Policy for a Multipolar World*, Maxwell Air Force Base, Ala.: Air University, School of Advanced Airpower Studies, 1992.

---

[8] References in this section were used to build the timeline in Figure 6.1 and are listed in order of publication. Undated references appear at the end of the section.

National Aeronautics and Space Administration, *NASA Management Instruction 1700.8: Policy to Limit Orbital Debris Generation*, April 1993.

National Aeronautics and Space Administration, *NASA Safety Standard: Guidelines and Assessment Procedures for Limiting Orbital Debris*, NSS 1740.14, August 1995.

Orbital Debris Program Office, "Orbital Debris Quarterly News, Volumes 1–14," Houston, Tex.: National Aeronautics and Space Administration, 1996–2010.

Scientific Advisory Board Ad Hoc Committee on Space Surveillance, Asteroids and Comets, and Space Debris, United States Air Force, *Report on Space Surveillance, Asteroids and Comets, and Space Debris; Volume 1: Space Surveillance*, SAB-TR-96-04, 1997.

Scientific Advisory Board Ad Hoc Committee on Space Surveillance, Asteroids and Comets, and Space Debris, United States Air Force, *Report on Space Surveillance, Asteroids and Comets, and Space Debris; Volume 2: Asteroids and Comets*, SAB-TR-96-04, 1997.

Scientific Advisory Board Ad Hoc Committee on Space Surveillance, Asteroids and Comets, and Space Debris, United States Air Force, *Report on Space Surveillance, Asteroids and Comets, and Space Debris; Volume 3: Space Debris Summary Report*, SAB-TR-96-04, 1997.

Ziegler, David W., *Safe Heavens: Military Strategy and Space Sanctuary Thought*, Maxwell Air Force Base, Ala.: Air University, School of Advanced Airpower Studies, 1997.

Limperis, Peter T., "Orbital Debris and the Spacefaring Nations: International Law Methods for Prevention and Reduction of Debris, and Liability Regimes for Damage Caused by Debris," *Arizona Journal of International and Comparative Law*, Vol. 15, No. 1, 1998.

Rex, Dietrich, "Will Space Run Out of Space? The Orbital Debris Problem and Its Mitigation," *Space Policy*, Vol. 14, No. 2, 1998, pp. 95–105.

Portree, Davis S. F., and Joseph P. Loftus, *Orbital Debris, a Chronology*, National Aeronautics and Space Administration, 1999.

Sarewitz, Daniel, Roger A. Pielke, Radford Byerly, Stanley A. Changnon, Rob Ravenscroft, Orrin H. Pilkey, Shirley Mattingly, Denis Walaker, Jack Fellows, J. Michael Pendleton, Ronald Brunner, and Thomas R. Stewart, *Prediction: Science, Decision Making, and the Future of Nature*, Washington, D.C.: Island Press, 2000.

Hyten, Colonel John E., "A Sea of Peace or a Theater of War? Dealing with the Inevitable Conflict in Space," *Air & Space Power Journal*, Fall 2002.

Prasad, M. Y. S., "Technical and Legal Issues Surrounding Space Debris—India's Position in the UN," *Space Policy*, Vol. 21, No. 4, 2005, pp. 243–249.

Johnson, Nicholas L., "The Historical Effectiveness of Space Debris Mitigation Measures," *International Space Review*, No. 11, December 2005.

United States Office of Science and Technology Policy, *U.S. National Space Policy*, Washington, D.C., 2006.

Johnson, Nicholas L., "Developments in Space Debris Mitigation Policy and Practices," *Proceedings of the Institution of Mechanical Engineers, Part G: Journal of Aerospace Engineering*, Vol. 221, No. 6, 2007, pp. 907–909.

Krisko, P., "The Predicted Growth of the Low-Earth Orbit Space Debris Environment—An Assessment of Future Risk for Spacecraft," *Proceedings of the Institution of Mechanical Engineers, Part G: Journal of Aerospace Engineering*, Vol. 221, No. 6, 2007, pp. 975–985.

Reilly, Caroline S., *Considering the Consequences of Space Warfare in the Geosynchronous Region*, London: King's College London, 2007.

Senechal, Thierry, *Orbital Debris: Drafting, Negotiating, Implementing a Convention*, Massachusetts Institute of Technology, 2007.

Carrico, Timothy, John Carrico, Lisa Policastri, and Mike Lou, *Investigating Orbital Debris Events Using Numerical Methods With Full Force Model Orbit Propagation*, American Astronautical Society, AAS 08-126, 2008.

Dillon, Matthew J., *Implications of the Chinese Anti-Satellite Test for the United States Navy Surface Forces*, Monterey, Calif.: Naval Postgraduate School, 2008.

Frey, Captain Adam E., "Defense of US Space Assets: A Legal Perspective," *Air & Space Power Journal*, Winter 2008.

Committee on the Peaceful Uses of Outer Space, United Nations, *Report of the Scientific and Technical Subcommittee on Its Forty-Sixth Session, Held in Vienna from 9 to 20 February 2009*, 2009.

"Flying Blind; Debris in Space—The Tragedy of the Commons Meets the Final Frontier," *Economist*, February 19, 2009.

Shaw, Colonel John E., "Guarding the High Ocean: Towards a New National-Security Space Strategy Through an Analysis of US Maritime Strategy," *Air & Space Power Journal*, Spring 2009.

Matthews, William, "Trackers of Orbiting Junk Sound Warning," *DefenseNews*, June 10, 2009. As of May 20, 2010:
http://www.defensenews.com/story.php?i=4133281

Orbital Debris Program Office, National Aeronautics and Space Administration, "Orbital Debris: Frequently Asked Questions," last updated July 7, 2009. As of May 27, 2010:
http://orbitaldebris.jsc.nasa.gov/faqs.html#3

"NASA: Debris No Threat to Space Station," *CNN*, December 1, 2009. As of September 19, 2010:
http://www.cnn.com/2009/TECH/space/12/01/space.station.debris/index.html

SpaceTech, Delft University of Technology, NASA-DARPA International Conference on Orbital Debris Removal, *Questionnaire on Space Debris Removal*, Chantilly, Va., December 8–10, 2009.

Johnson, Nicholas L., and Eugene G. Stansbery, "The New NASA Orbital Debris Mitigation Procedural Requirements and Standards," *Acta Astronautica,* Vol. 66, No. 3–4, February–March 2010, pp. 362–367.

Johnson, Nicholas L., "Orbital Debris: The Growing Threat to Space Operations," paper presented at 33rd Annual AAS Guidance and Control Conference, Breckenridge, Colo., American Astronautical Society, February 6–10, 2010.

Marks, Paul, "Nanosatellite Sets Sail to Tackle Space Junk," *New Scientist,* March 26, 2010. As of September 19, 2010:
http://www.newscientist.com/article/
dn18705-nanosatellite-sets-sail-to-tackle-space-junk.html

Kehler, General Robert, USAF, "Air Force Space Command 'On the Edge'," *2010 National Space Symposium,* Colorado Springs, Colo., April 13, 2010.

de Selding, Peter B., "Europe Keeping Increasingly Capable Eye on Orbital Debris," *Space News,* April 21, 2010. As of April 25, 2010:
http://www.spacenews.com/civil/100421-europe-eye-orbital-debris.html

Hsu, Jeremy, "Solar Sails Could Clean Up Space Junk," space.com, April 28, 2010. As of April 29, 2010:
http://www.space.com/businesstechnology/solar-sails-clean-space-junk-100428.html

Gross, Doug, "Rogue Satellite May Impact Cable TV in U.S.," *CNN,* May 12, 2010. As of May 12, 2010:
http://www.cnn.com/2010/TECH/ptech/05/12/rogue.satellite.tv/index.html

Telephone conversation with Nicholas L. Johnson, Chief Scientist for Orbital Debris, National Aeronautics and Space Administration, June 1, 2010.

Jones, Johanna, Defense Advanced Research Projects Agency, "Fact Sheet: DARPA Conducting Study on Orbital Debris Removal," undated. As of June 15, 2010:
http://www.darpa.mil/Docs/OrbitalDebris.pdf

Orbital Debris Program Office, National Aeronautics and Space Administration, "Orbital Debris Mitigation," undated. As of May 27, 2010:
http://orbitaldebris.jsc.nasa.gov/mitigate/mitigation.html

Orbital Debris Program Office, National Aeronautics and Space Administration, "Orbital Debris Mitigation Standard Practices," undated. As of May 27, 2010:
http://orbitaldebris.jsc.nasa.gov/library/USG_OD_Standard_Practices.pdf

Space Track home page, "Space Track: The Source for Space Surveillance Data," undated. As of June 11, 2010: http://www.space-track.org/perl/login.pl

Radon[9]

"Radon Safety Standard Issued," *Facts On File World News Digest*, August 22, 1986.

"Radon Cleanup Bills Introduced," *Facts On File World News Digest*, March 20, 1987.

"Radon Threat Found Widespread," *Facts On File World News Digest*, August 7, 1987.

"Radon Risk Stressed," *Facts On File World News Digest*, September 23, 1988.

"Radon Detection Bill Enacted," *Facts On File World News Digest*, November 4, 1988.

"School Radon Tests Urged," *Facts On File World News Digest*, April 28, 1989.

"Reduced Danger Seen in Radon," *Facts On File World News Digest*, March 21, 1991.

United States General Accounting Office, *Air Pollution: Actions to Promote Radon Testing*, Washington, D.C., GAO/RCED-93-20, December 1992.

"EPA Issues Radon Recommendations," *Facts On File World News Digest*, August 29, 1993.

"Study Probes Radon Dangers," *Facts On File World News Digest*, September 19, 1996.

"Panel Links Radon to Lung Cancer," *Facts On File World News Digest*, February 26, 1998.

United States General Accounting Office, *Drinking Water Revisions to EPA's Cost Analysis for the Radon Rule Would Improve Its Credibility and Usefulness*, Washington, D.C., GAO-02-333, February 2002.

Tiemann, Mary, *Safe Drinking Water Act: Implementation and Issues*, Congressional Research Service, August 20, 2003.

Tiemann, Mary, *Drinking Water State Revolving Fund (DWSRF): Program Overview and Issues*, Congressional Research Service, October 24, 2008.

---

[9]   References in this section were used to build the timeline in Figure 6.1 and are listed in order of publication. Undated references appear at the end of the section.

United States Environmental Protection Agency, *A Citizen's Guide to Radon: The Guide To Protecting Yourself And Your Family From Radon*, EPA 402/K-09/001, January 2009.

United States Environmental Protection Agency, *Consumer's Guide To Radon Reduction: How to fix your home*, EPA 402/K-10/002, January 2010.

Spam[10]

"FTC Cracks Down on 'Spam'," *Facts On File World News Digest*, April 4, 2002.

"Efforts to Cut Down on E-Mail 'Spam'," *Facts On File World News Digest*, November 8, 2002.

"Special Report: Spam Turns 25," *Facts On File World News Digest*, May 2003.

"Senate Panel Votes to Limit 'Spam'," *Facts On File World News Digest*, July 10, 2003.

"New York Charges 'Spammer'," *Facts On File World News Digest*, July 17, 2003.

"Senate Passes Antispam Bill," *Facts On File World News Digest*, November 6, 2003.

"Bush Signs Antispam Law," *Facts On File World News Digest*, December 25, 2003.

"'Spam' E-Mail Lawsuits Filed," *Facts On File World News Digest*, December 31, 2004.

McDowell, Mindi, and Allen Householder, "National Cyber Alert System Cyber Security Tip ST04-007: Reducing Spam," last updated July 29, 2009. As of March 26, 2010:
http://www.us-cert.gov/cas/tips/ST04-007.html

Consumer & Governmental Affairs Bureau, Federal Communications Commission, "CAN-SPAM: Unwanted Text Messages and E-Mail on Wireless Phones and Other Mobile Devices," August 11, 2009. As of March 1, 2010:
http://www.fcc.gov/cgb/consumerfacts/canspam.html

---

[10] References in this section were used to build the timeline in Figure 6.1 and are listed in order of publication. Undated references appear at the end of the section.

U.S. Border Control[11]

"Border Corruption Cited," *Facts On File World News Digest*, September 1, 1973.

"Electronic U.S.-Mexican Border Barrier," *Facts On File World News Digest*, September 1, 1973.

"U.S. Plans Border Fence," *Facts On File World News Digest*, October 27, 1978.

"Record Border Crossings Reported," *Facts On File World News Digest*, February 23, 1979.

"U.S. Drops Border Fence," *Facts On File World News Digest*, March 4, 1979.

"Illegal Border Crossings Rise," *Facts On File World News Digest*, March 7, 1986.

"Mexican Border Ditch Proposed," *Facts On File World News Digest*, February 3, 1989.

United States General Accounting Office, *Survey of U.S. Border Infrastructure Needs*, Washington, D.C., GAO-NSLAD-92-56, November 1991.

"Mexican Border Strengthened," *Facts On File World News Digest*, January 25, 1996.

United States General Accounting Office, *U.S.-Mexico Border Issues and Challenges Confronting the United States and Mexico*, Washington, D.C., GAO/NSIAD-99-190, July 1999.

"Man With Bomb Parts Caught at U.S. Border," *Facts On File World News Digest*, December 23, 1999.

"U.S. Border Security Pact Signed," *Facts On File World News Digest*, June 15, 2000.

"Senate Passes Border Security Bill," *Facts On File World News Digest*, April 25, 2002.

"Bush Signs Border Security Bill," *Facts On File World News Digest*, May 30, 2002.

"House Passes Border Security Bill," *Facts On File World News Digest*, February 10, 2005.

"New Border Passport Rules Announced," *Facts On File World News Digest*, April 21, 2005.

"Volunteers Launch Patrols on Arizona Border," *Facts On File World News Digest*, April 21, 2005.

---

[11] References in this section were used to build the timeline in Figure 6.1 and are listed in order of publication. Undated references appear at the end of the section.

"New Mexico, Arizona Declare Emergencies," *Facts On File World News Digest*, September 1, 2005.

"House Passes Border Security Bill," *Facts On File World News Digest*, December 31, 2005.

"House Passes Border-Fence Bill," *Facts On File World News Digest*, September 28, 2006.

"Senate Clears Border Fence Bill," *Facts On File World News Digest*, October 5, 2006.

"Bush Signs Border Fence Bill," *Facts On File World News Digest*, November 2, 2006.

United States Government Accountability Office, *US-VISIT Program Faces Strategic, Operational, and Technological Challenges at Land Ports of Entry*, Washington, D.C., GAO-07-248, December 2006.

"'Virtual Fence' Delayed Due to Pilot Flaws," *Facts On File World News Digest*, March 13, 2008.

"Border Fence Project Delayed," *Facts On File World News Digest*, September 25, 2008.

Nuñez-Neto, Blas, *Border Security: The Role of the U.S. Border Patrol*, Congressional Research Service, November 20, 2008.

United States Customs and Border Protection, Department of Homeland Security, "Secure Border Initiative," November 24, 2008. As of February 19, 2010: http://www.cbp.gov/xp/cgov/about/organization/comm_staff_off/mark_ borkowski.xml

"Obama to Review Border Troop Request," *Facts On File World News Digest*, March 13, 2009.

"U.S. Increases Mexico Border Security," *Facts On File World News Digest*, March 26, 2009.

Kim, Yule, *Protecting the U.S. Perimeter: Border Searches Under the Fourth Amendment*, Congressional Research Service, June 29, 2009.

United States Government Accountability Office, *Secure Border Initiative Technology Deployment Delays Persist and the Impact of Border Fencing Has Not Been Assessed*, Washington, D.C., GAO-09-896, September 2009.

Haddal, Chad C., *Border Security: Key Agencies and Their Missions*, Congressional Research Service, January 26, 2010.

United States Department of Homeland Security, "U.S. Customs and Border Protection Timeline," undated. As of April 2, 2010: http://nemo.customs.gov/opa/TimeLine_062409.swf

Index